P9-CLU-638

MY REMARKABLE JOURNEY

MY REMARKABLE JOURNEY

A MEMOIR

KATHERINE JOHNSON

with Joylette Hylick and Katherine Moore

and with

Lisa Frazier Page

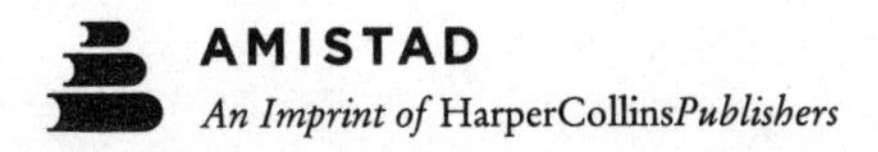

MY REMARKABLE JOURNEY. Copyright © 2021 by Katherine Johnson. All rights reserved. Printed in the United States of America. No part of this book may be used or reproduced in any manner whatsoever without written permission except in the case of brief quotations embodied in critical articles and reviews. For information, address HarperCollins Publishers, 195 Broadway, New York, NY 10007.

HarperCollins books may be purchased for educational, business, or sales promotional use. For information, please email the Special Markets Department at SPsales@harpercollins.com.

First Amistad hardcover published 2021

FIRST EDITION

Designed by THE COSMIC LION

Library of Congress Cataloging-in-Publication Data has been applied for.

ISBN 978-0-06-289766-4

21 22 23 24 25 LSC 10 9 8 7 6 5 4 3 2 1

To my husband and parents, who were always a great source of inspiration and support, and to young people seeking to make a better world through the field of science

CONTENTS

Foreword by Dr. Yvonne Darlene Cagle ix
Introduction: An Unimaginable Century 1

1 Nobody Else Is Better Than You 15
2 Education Matters 31
3 A Time for Everything 55
4 The Blessing of Help 71
5 Be Ready 93
6 Ask Brave Questions 107
7 Tomorrow Comes 125
8 Love What You Do 143
9 Shoot for the Moon 165
10 Finish Strong 185
11 Land on a Firm Foundation 199

Acknowledgments 217
Notes 221

FOREWORD

I knew about Dr. Katherine Johnson more than two decades ago, but I didn't know I knew her. A few years ago, in 2016, I went to the first reunion of black astronauts. During one part of the evening, Joylette Hylick, Dr. Johnson's daughter, was a speaker, and she was telling this amazing story about her mother. I wanted to meet Dr. Johnson. Everybody did. And so I tried to get to Joylette after the talk, but I couldn't. Just about the time that I was ready to give up, and she was about to walk away, she turned to me and said, "Dr. Cagle?"

And I said, "Yes, I have this picture that I signed for your mother." I told her that even if I'm never able to meet her, it would mean everything if she would take this picture to her. Joylette looked at me said, "Dr. Cagle, you've already met mom."

And I was thinking, no, I would have known. I thought she must have confused me with someone else. But Joylette insisted, "No, you have met mom! As a matter of fact, you've taken a picture with her." She saw the doubt in my mind and knew I wasn't fully convinced yet, so she said, "I'll tell you what. I'm going to prove it to you."

About two weeks later, I got a text message, and it said, "I told you so!" I opened it, and the first thing I saw was this picture of me in 1997, decades ago, and I was blown away. And all of a sudden my eyes dropped down, and I saw this woman who looked very familiar. It was an awards picture, and I was one of the award recipients. Sitting directly in front of me . . . oh my gosh, when it hit me, I was just floored. It was Dr. Katherine Johnson! And it all came back to me. It had been my first astronaut appearance ever. I had just finished my astronaut candidacy training, and this was the first event they allowed me to attend. It was in Philadelphia for the National Technology Association's annual convention. The NTA was established as a black engineering association before black engineers were allowed to join other groups. Dr. Johnson at one time had been the organization's treasurer. There was an awards ceremony later that night, and the organization recognized a few people, including Dr. Johnson and me.

I remember the presenter mentioning an African American woman, an African American mathematician from NASA, and I remember thinking, why don't we know about each other? I was in my blue astronaut suit, and I remember thinking that we need to connect. I wanted to go afterwards to meet her, but there were so many people surrounding the astronaut blue suit that I never got to her. I figured we would meet somewhere again, but it never happened—at least not for nearly two decades.

When I met Dr. Johnson face to face a short time after that encounter with her daughter, it was such a seminal moment for me. But one of the things I have a hard time reconciling is knowing there was so much lost time. This woman became my real heroine. She could've been my mentor years before. Who I would have been and what I could have done under her mentorship, I just think was a gift that I can't regain. But it makes the time that I had with her that much more precious and impactful.

It begs the question: why didn't I know about her even before 1997? Why didn't all little girls know, especially little girls of color? Why did I have to go through so much hurt and heartache in life when I could have looked at her and held my head high and stepped through with such poise, comportment, and grace? Why couldn't I have had that voice that spoke, not just to the world, but to my heart, to my resolve? I cannot find consolation, reconciliation, or any rational reason why that had to be so.

That's why after I did get to know her, she became more than a mission in my mind. She became a movement, an entire movement. I got to know her and her story. And I was more present than I might have been if I had not felt the headwind, the cold, the sorts of barriers I had to encounter on my own. As I learned about her story, it became my story, and we became one. Her words became my words. Her voice largely became my voice. I actually found my voice in sitting at her feet, learning from her, and hearing her voice about how she stepped through it. When she stepped through the barriers, her steps were hidden. They were silent. But when she was given voice, she spoke about it with such uncompromising grace. Whether there were barriers or bands or badges, it didn't matter to her. It was all the same. She was going to do her best. She was just doing her job. And that's what she said, "Always do your best."

The family invited me to go to NASA's screening at Langley of the movie, *Hidden Figures*, and there were many other events. We ended up going to the New York screening, the Smithsonian, and the White House. We did a number of interviews together. But to me, you take away the red carpet and the cameras, and it was always just she and I having a conversation. After about a year or two, that eventually turned into our Sunday conversations, Sunday chats. We just continued to have these Sunday chats back and forth, where we let our hair down and talked about absolutely everything.

We started out talking about the science, the math, and the engineering, but there came a point where we talked about everything from her life, my life. I know what it's like to be that hidden figure. That's why it affects me so deeply. She never made me feel hidden. That's why those Sundays were so important.

When asked how she dealt with the discrimination and segregation at NASA during her era, she said she was always aware of it but that when she passed through those gates, she had a job to do. And no matter what was going on, she always did her best. She didn't have perspective on the impact of what she did. And so she genuinely didn't understand all the attention she was getting after the book and the movie. "I was just doing my job," she said. The way she saw it, she was doing the work that the men wouldn't do because they were doing more important things.

In 2018, the whole world was celebrating when it was announced that NASA was planning to go back to the moon. It had been several weeks, and no one from NASA had thought to tell Dr. Mom (that's what I began calling her), and so I told Joylette that I would love the honor of letting her know we were going back. When I told her over the phone, I was sitting at my kitchen table in California, and she was in Virginia. And she was so excited. She said, "That's just wonderful. Now, this is what you need to do." She immediately went to work. She said, "The first thing you need to do is plant a garden." I thought I had misheard. In my mind, I'm thinking it's a wonderful, beautiful idea, but at one hundred years old, you can envision whatever you want. Turns out, she knew exactly what she was talking about. Right after that, she said, "You can't just plant a garden. You have to make sure the soil content is correct because if you don't have enough nitrogen in the soil, nothing is going to grow, and it's primarily carbon dioxide you've got to look at. You've got to figure out how to make the carbon dioxide-oxygen ratio just right so you can grow the plants, and the plants can give off the oxygen." Then I realized she was engineering, and she was ahead of me. I needed to get on board and catch up to her. And so instead of trying to question, correct, or whatever I might want to do, I got a piece of paper and started writing furiously. And thirty-five minutes later, I had a recipe for planting a garden on the moon. I knew it was precise and correct because I am a biochemist and a physician, and I knew those are the compounds that you need to make sure are in the right concentrations and ratios. I was just writing and thinking, "One hundred years old, I can't believe it." Her mind was so sharp all the way to the day before she passed on. I say passed on, not passed away, because she is still with us.

I rarely give a speech now without mentioning her. In all my speeches, I always say the launch speed of a space craft is 17,500 miles per hour. There's a scene in the *Hidden Figures* movie where the character playing Dr. Johnson (Taraji P. Henson) is writing on the board what that velocity needs to be for the space vehicle to make it into orbit. So that number, 17,500 miles per hour, was emblazoned in my mind. I was so impressed that the movie got it right. And then in that scene Taraji keeps writing out many, many more decimal places. And that's just how it was with Dr. Johnson. Even if I got the right answer or represented something correctly, I always had to catch up to her because she could take it many more decimal points out.

I asked her what she thought of the movie. She said the movie was wonderful, but they got two things wrong. First of all, she didn't wear glasses until much later in life, and secondly, she said the movie made it seem like she was really anxious at that first moon launch. Well, she had already launched Alan Shepard. She had orbited John Glenn. And then there was the moon landing, and she knew her numbers were right.

She also put the first antennae up in space, and she continued to work with NASA even into projecting going into Mars. And she did so much more that is never talked about. I know this because it is written in an old NASA evaluation. The first time I flew to Virginia to meet her, Joylette and I were talking in her room until two in the morning and going through all these pictures and things of her history. I wanted to know everything. There were so many awards and papers, and Joylette was trying to figure out what to keep. She pulled out this one thing. It was a full sheet of paper, and she wasn't sure what it was. I looked at it and start reading it, and it was a NASA work evaluation of Dr. Johnson from 1986. There was a whole page about her work. I read every line, and as I was reading these things, I was blown away. That evaluation was pure gold.

So, when it was time for the Oscars, the producers called me after talking to the family and said they'd like to have Katherine Johnson at the Oscars since *Hidden Figures* had been nominated in a number of categories, including "Best Picture." She was ninety-eight years old at the time, and I was thinking most ninety-eight year olds don't even get in a cab by themselves, and they're talking about putting her on a plane and flying her across the country. As a physician, I knew what the answer was. But I also knew Dr. Katherine Johnson. She always wanted to speak for herself. And so I said to them—I didn't want to be the spoiler of the party—I said, as a physician, there were some things we had to take into consideration, but I couldn't speak for Dr. Johnson and that they would have to speak to her. I knew she would want that. They said, "Well, will you ask her and see what she says?" I talked to Joylette and Kathy, her youngest daughter, and they talked to Dr. Mom, and her response was, "You're not going to do this without me!"

Well then, I was on the hook. I was counting on her to say it was too far and too much. I knew she was going to need a medical person on the plane, a physician, because of her age and the travel. I knew it had to be somebody who knew aviation medicine and someone who knew her well enough to recognize early on if she was in any kind of distress. I knew I had to do it, even though I knew what a daunting task it would be and even though I knew that if she had asked me, I would have probably discouraged her. But I had turned the option over to her, and she had made the call. I couldn't think of anyone else I could recommend who met all of the criteria and would make her the priority. I told the family and the producers that I wanted to do it. It was a journey of love, and I brought my "A game." I flew in my flight suit because I was a flight surgeon for twenty-two years.

It was such an honor to have the opportunity to travel with her and push her onto the Oscars stage in her wheelchair. On the night of the show, the people coordinating the event had told me that the presentations were on a tight schedule. But they said this was such a historic moment, once in a lifetime really, so when the applause started, I shouldn't step back until it was finished. They told me to let the world recognize her and applaud her for as long as they wanted. And so when the applause started, I didn't know anyone was going to stand. And they rose, and they rose all the way up, four tiers up. Everyone was standing. Everyone was crying. The applause was resounding, and they weren't stopping. And finally it started to settle down a little bit, and I started backing away slowly. But I wasn't sure whether to leave or stay. Denzel Washington was standing right there, and our eyes met, and it was almost as if he knew what I was struggling with. He just gave me the two fingers, like come back for an encore, and they wouldn't stop applauding. I backed away again, and he nodded, bring her back. And they applauded for a while longer. I was so happy. That was the right thing to do.

So afterwards, I get her back to the room, she's in bed, and we're just talking. She was wide awake. And a question came to my mind.

"Dr. Mom, I have a question for you," I said to her. "If it had been possible, did you ever think about going to space yourself? You being an astronaut?"

Her eyes lit up, and she said, "Ohhh, baby! I would have loved to have gone into space myself!"

In that moment, my heart broke. I didn't want her to see the tears, but everything just shattered inside of me. It's one thing to do everything she did to land a human on the moon. It's a completely different conversation if somewhere deep in your heart you always wanted that person to be you, and you couldn't even aspire to it. You could dream, but that's as far as it could go. She had carried that dream with her all that time, from a little girl counting the stars, and here she was still with that yearning.

As an African American woman who *could* aspire to travel to the moon, I am the embodiment of her dream. And to all those African American girls who will walk on the moon and land on Mars someday, we are her legacy.

Dr. Yvonne Darlene Cagle is an astronaut for the National Aeronautics and Space Administration (NASA), family physician, and retired colonel in the United States Air Force, where she served as a senior flight surgeon. She has served as a professor at a number of universities, including Stanford, University of California Davis, and University of Texas Medical Branch in Galveston.

MY REMARKABLE JOURNEY

introduction

AN UNIMAGINABLE CENTURY

One hundred years. Who really expects to live long enough to see an entire century? Few people in this country—less than 1 percent—make it that far. So, one day in 2018, as my centennial birthday approached, I made light of the moment with my oldest daughter, Joylette.

"If I knew I'd live this long, I'd have taken better care of myself," I whispered, quoting to her a line I'd heard somewhere before.

We both laughed.

The truth is I'm not complaining. I've had a wonderful life. Every day that I wake up in my right mind is a blessing, and I've been richly blessed to see the world change in miraculous ways. Think about it. I've been around longer than sliced bread, which didn't become one of the century's great inventions (or at least the thing by which everything good is compared) until 1928. By then I was already ten. When I was born on August 26, 1918, in White Sulphur Springs, West Virginia, the world was a very different place. The First World War was ending, and as soldiers made their way back to their families, a mysterious flu was spreading across the world in a pandemic that would claim an estimated 50 million lives over the next two years; a total of 675,000 people died in the United States of the virus, which came to be known as the Spanish flu. The year I was born Ford Motor Company also was selling its popular Model T for about $350. That price tag made cars suddenly affordable for the first time to many middle-class Americans. But the all-black, mass-produced vehicles were still out of reach for most poor and working-class families, particularly in rural communities, like the one where I grew up in the Allegheny Mountains of West Virginia. Of course, I was too young to know any of this firsthand, but I remember my father transporting our family around throughout my childhood in a horse-drawn buggy. In those years, most of what we needed was within walking distance. When we couldn't walk, Daddy hitched the horses. My parents never owned a car.

I couldn't have imagined as a child that someday highways would stretch from one side of the country to the next, zigzagging through the American landscape in all directions. Or that someday vehicles of all colors, shapes, and sizes would fill those roadways with busy people, always in a hurry to get to the next important place. Or that someday vehicles would park themselves. I was born at a time when women couldn't vote, my people were called colored and treated as second-class citizens, and white lynch mobs terrorized our communities throughout the nation, particularly in the South. So even in my most vivid imagination—and I had a pretty creative one as a child—I could not envision the life I would live.

I have lived through eighteen US presidents and mourned with the nation when two of the great ones died tragically in office. Never have I been prouder of this country than when we elected Barack Obama as the first African American president of the United States, number forty-four. My husband, Jim, and I had been so hopeful that we sent a financial contribution to the Obama campaign. Then, on election night in 2008, we watched in astonishment as the poll results rolled in, confirming a victory we never thought we'd witness in our lifetime. America had looked beyond race and perhaps finally was ready to move past the racial fear, hatred, and misunderstanding that had bitterly divided the nation my entire life. I don't spend much time looking back or dwelling on race, but that night I couldn't help thinking about my father, who had grown up in the shadows of slavery and had been limited by his race to a sixth-grade education. Yet Daddy recognized the value of education and made great sacrifices to assure that my sister, two brothers, and I were able to graduate from college. Daddy would have been so proud of the articulate, progressive, well-educated African American man and woman about to occupy the White House.

My older brother, Charlie, ninety-one years old in 2008, also was a big Barack Obama fan. One of his proudest possessions was an Obama cap that Kathy, my youngest daughter, had bought him. Charlie wore that cap everywhere. He was so eager to vote for Obama that he cast his vote early from a hospital bed via absentee ballot. Then, on November 3, 2008, just one day before the election, Charlie passed away. It brought me solace knowing that my brother's vote had helped to make history.

So it was one of the greatest honors of my life when President Obama invited me to the White House in November 2015 to receive the Presidential Medal of Freedom, the nation's highest civilian award. At ninety-seven years old I was being honored for my work as a research mathematician at NASA and its predecessor from 1953 to 1986. My first thought was that I couldn't believe it. What had I done to deserve such an esteemed award? Then I wished I could have shared the honor with my team. Though I was eventually assigned to a specific research team, I had started at the National Advisory Committee for Aeronautics (NACA), the predecessor to NASA, as part of the West Area Computing Unit, the segregated section of black female mathematicians at the agency's Langley Research Center in Hampton, Virginia. In the days before integration and before electronic computers, we worked as human computers, doing the mathematical calculations manually for the space program. The women in that unit were some of the smartest people I've ever known. They included Dorothy Vaughan, who supervised the "West Computers," as we were called, and was one of the first black supervisors at NASA; Eunice Smith, my good friend and neighbor, who had been a mathematician there nine years by the time I arrived; and Mary Jackson, the agency's first black female engineer. Mary passed away in 2005 at age eighty-three; Eunice died a year later, at eighty-two; and Dot, in 2008 at ninety-eight. All three of the women enjoyed good, long lives, but they never got to experience the widespread public admiration that would come years later when the world discovered the important role that African American women played in helping to send the first humans to the moon. I don't know what any of them would have made of all the attention, but I certainly never expected any glory. I was just doing the job I was hired to do.

Nevertheless, I was extremely grateful and humbled to meet President and Mrs. Obama and to receive such an honor. I still could hardly believe my eyes when I was ushered into the elegant East Room of the White House for the ceremony and seated right next to baseball legend Willie Mays, a fellow award winner. The quiet room seemed to grow even more still when President Obama began speaking, welcoming everyone to the White House. And then he said my name:

> Growing up in West Virginia, Katherine Johnson counted everything. She counted steps. She counted dishes. She counted the distance to the church. By ten years old, she was in high school. By eighteen she had graduated from college with degrees in math and French. As an African American woman, job options were limited, but she was eventually hired as one of several female mathematicians for the agency that would become NASA. Katherine calculated the flight path for America's first mission in space, and the path that put Neil Armstrong on the moon. She was even asked to double-check the computer's math on John Glenn's orbit around the Earth.

The irony made people laugh. President Obama continued in a jovial mood:

> So if you think your job is pressure-packed, hers meant that forgetting to carry the one might send somebody floating off into the solar system. In her thirty-three years at NASA, Katherine was a pioneer who broke the barriers of race and gender, showing generations of young people that everyone can excel in math and science and reach for the stars.

A few minutes later, President Obama walked over to me, draped the medallion around my neck, and kissed me on the cheek. "Thank you," I whispered, looking up to him, as the audience applauded. Still seated, I turned my face to the guests, nodded my head, and mouthed again, "Thank you, thank you." In the years since that wonderful day, so many people, especially the women, have asked me how it felt to be kissed by President Obama. All I can say is it was thrilling.

Little did I know that the White House ceremony was just the first big wave in an avalanche of attention headed my way. The next year, on May 6, 2016, NASA did the craziest thing—named a new, forty-thousand-square-foot building at Langley in my honor! The agency held a ceremony to unveil the name, and I attended, along with my family, friends, NASA employees, and other guests, totaling at least three hundred people. I'd spent so much of my life on these grounds and was incredibly moved—and again truly humbled—to see my name stretched across the front of this beautiful new building, the Katherine G. Johnson Computational Research Facility. It consolidates the agency's data in one place and provides lots of new office space. As long as this building stands, it will bear my name. Truly unbelievable. As I sat there, listening to the kind words from NASA officials and the local and state politicians, I wished again that my research team and all of the women who worked as computers could be sharing in this recognition. I was always proud of my work, but for Pete's sake, I didn't do anything alone.

The keynote speaker for the unveiling ceremony was Margot Lee Shetterly, who wrote the book that told our story to the world: *Hidden Figures: The American Dream and the Untold Story of the Black Women Mathematicians Who Helped Win the Space Race.* Margot is a lovely young woman whose father earned prominence at Langley as a climate scientist. She had grown up in Hampton, Virginia, and knew many of the West Computers through family and community connections. She spent many hours over a few years, interviewing me for the book, as well as dozens of others, but I had no idea of the impact this hidden history would have on the world once it was revealed.

The book was released in September 2016 and quickly became a bestseller. The movie *Hidden Figures*, based on the book, was released in January of the following year. To prepare for the role, Taraji P. Henson, the beautiful actress who played me in the movie, had traveled to Newport News and spent several hours with me in my home. She was very kind, down-to-earth, and curious, and I enjoyed our time together. I saw the movie at least three times, and I must admit it was quite a strange feeling to hear my name and see someone else walking in my shoes. But Taraji and all of the actors did a phenomenal job. Remember, though, it's a movie, so not every detail was true to my story. I'd estimate that about 75 percent of what was shown in the film is accurate. Some scenes were embellished or made up to increase the drama. For example, Dot, Mary, and I did not commute to work together. I mostly rode with Eunice Smith, who was mentioned in the book but wasn't a character in the movie. Also, I never had to rush back and forth across the Langley campus to use a segregated bathroom. While there were indeed separate bathrooms for white and black employees for much of my time at Langley, I always used the one closest to my workspace. At first, I didn't even realize there were separate bathrooms because only the "colored" bathrooms were marked. There was no sign outside the restrooms designated for my white counterparts, and some buildings didn't have a "colored" bathroom. So there were humiliating times when black employees had to walk a long distance to a different building to find a designated bathroom to relieve themselves. And that's the ridiculous reality the movie depicted through my character. But in real life, I didn't follow the rules. I figured I was as good as anybody else, so even after I realized there were "colored" restrooms, I just refused to use them.

Was I able to get away with it because of my fair skin complexion? That's probable. I'd overheard a few conversations that let me know some of my white counterparts weren't quite sure of my race. But as I saw it, that was their issue. I never tried to hide who I was. When people talk to me about the movie, many of them ask about the bathroom scene, and they seem surprised when I tell them the truth. The truth was more complicated, and it is sometimes difficult for a movie to explain such nuances. Also, I learned that some of the characters in the film were what the industry calls a "composite," meaning the single role was based on the characteristics of multiple people. At least a couple of the characters—supervisor Al Harrison (played by Kevin Costner) and Paul Stafford (played by Jim Parsons) were composites of a number of my white NASA colleagues. The fictional character Vivian Mitchell (played by Kirsten Dunst) was created to show the condescending attitude that some white female managers displayed toward their usually better-educated black female counterparts. Overall, I enjoyed the movie and was happy for the cast members when the film was nominated in the "Best Picture" category. What happened next, though, was another of those times that went beyond my imagination: I was invited to participate in the 2017 Academy Awards Show (the Oscars).

When the film company first called to ask me to attend, my daughters, Joylette and Kathy, said no. At my age, I would need far too many special accommodations to travel from the East to the West Coast safely. But the film company representatives persisted, asking what might make the trip possible for me. Certain that the requirements would be too much, my girls started listing the necessities: a private plane that could change course, if needed, in an emergency, a doctor to travel with me to be able to respond immediately in an emergency, twenty-four-hour nursing assistant care; and I'd have to have accommodations for at least two days before the trip and two days afterward so I could be well rested. My girls were taking no chances with my health. They also referred the film company's representatives to Dr. Yvonne Cagle, a NASA astronaut, medical doctor, and family friend. The film company didn't flinch. They agreed to all of our requests and more, including a private hotel suite for me, four additional rooms for my family, tuxedos for the guys, and a so-called glam squad for the ladies. They must have *really* wanted me there.

My daughters accompanied me to Los Angeles for a weeklong trip, as well as four of my grandchildren—Troy, Laurie, Michele, and Michael. Two of my grandsons, Greg and Doug, were not able to make it, and "the G-6," as our family lovingly calls my grandchildren, didn't feel quite whole without them. Dr. Cagle, a highly regarded medical doctor and senior flight surgeon, was able to travel with us as the doctor. She is a brilliant African American woman, retired US Air Force colonel, and visiting Stanford University professor who is making her own research history at NASA. The two of us have bonded over the years, and she surprised me with news that her celebrity clothing designer, Angela Dean, who has dressed a long list of superstars (from Nancy Wilson and Patti LaBelle to Madonna) wanted to make me a special dress for the red-carpet event. My daughters and I were honored by the designer's generosity, and when she presented the dress to us, our jaws dropped. She had taken such meticulous care to design a dress that not only was stunning but also very practical. She chose periwinkle blue because her research showed it was a popular color during my NASA era. She added crystals to the rounded neckline and sleeves so I wouldn't need much jewelry. She tapered the wrists to keep the sleeves in place when I raised my arm to wave. And she lined the dress in a super soft fabric to keep me warm and comfortable. She literally thought of everything, and the dress made me feel ready to meet Hollywood.

On the night of the show, which was televised around the world, the three costars of *Hidden Figures* stood together onstage in their beautiful evening gowns and perfect hair and faces. Taraji introduced me, and I heard her call me a "true NASA and American hero." Yvonne, in her own exquisite gown, pushed me in a wheelchair to center stage. I may have looked calm, but my heart was beating so fast. I smiled and looked out into the audience, but I couldn't see individual faces because of the bright lights. But I saw a collage of men in tuxedos and women in glittering gowns, filling nearly every space of that huge, multitier theater. And they were standing and applauding. All of this for me? My ninety-eight-year-old heart could barely contain the gratitude. It was too much. Finally, with some prompting from Yvonne, I managed to say four little words: "Thank you very much!"

Though there had been many West Computers, the movie focused primarily on the lives of Dot, Mary, and me. With the two of them in heaven, the publicity spotlight turned to me. The wider world wanted to know more about the hidden figure. I hadn't been hidden from my Hampton/Newport News community—my church members at Carver Memorial Presbyterian, my sisters from Alpha Kappa Alpha Sorority, Inc., my friends on the faculty and staff at Hampton University, or my many neighbors through the years. But the invitations suddenly began pouring in for me to speak, attend events, accept awards, and more. At my age I was no longer able to travel long distances, and so those duties fell primarily to my daughters and grandchildren, who seemed to be traveling to a different city every week. But there was one invitation I had to accept, one journey I was most eager to take. For my ninety-ninth birthday, the library in my hometown, White Sulphur Springs, West Virginia, was being renamed in my honor, and officials asked if I could attend the ceremony.

My family decided to make a mini reunion of the weekend, and we took some time before the ceremony to drive down Church

Street, past some of the long-gone places I'd loved as a child: the Big House, the two-story white house where my family once lived; the Core family's home across the street; the Mary McLeod Bethune Grade School for colored children, which surprisingly is now privately owned and has been redesigned as a private residence; the colored cemetery, where many of my family members are buried; and the Methodist and Baptist churches, where we'd attended Sunday services and Bible school in the summer. I watched through the car windows and eagerly pointed out all the familiar places. My daughters were surprised that I remembered every name. Once we made our way to the library renaming ceremony, I learned there was yet another surprise for me. The library was dedicating a large room inside the building in honor of my father, who had once worked there as a custodian. The room was to be called the Joshua McKinley Coleman Community Room. Daddy would have loved it. He had been denied the full education he desired in his youth, but now the community gathering spot inside this historical place of learning would carry his name for generations to come. I could only hope that from time to time, the most curious among the congregants would wonder about the man behind the name on the door and get to discover how truly great he was.

For my one hundredth-birthday weekend, I returned home again—this time to West Virginia State University, my alma mater—for another special honor. The university unveiled a beautiful, life-size bronze statue of me and created an endowed scholarship in my name. The scholarship will assist students who are studying in the STEM (Science, Technology, Engineering, and Math) fields, and I hope both will inspire a love of learning and a love of math in young people long after I am gone. That kind of inspiration was what I found when I first stepped onto this campus as a freshman who'd just turned fifteen years old. We were very proud that with two PhDs on staff, the math department at West Virginia State was second to none. When a professor told me one day a couple of years later that I'd make a good research mathematician, he gave me a goal to pursue. It was the 1930s, and I didn't have a clue what a research mathematician was. But it didn't matter that I was a Negro or that I was a woman. My professor believed I'd make a good research mathematician, and so I had no doubt I could figure it out someday and become one. Now, standing on this campus, forever in bronze is proof that I did just that. There was no better way to celebrate a hundred years.

Since then, the accolades have kept coming, and I've done my best to keep living. A reporter once asked me what I thought was my greatest accomplishment. She probably thought I'd say something about my contributions to NASA or the field of math, but I responded without hesitation, "Staying alive." No one has been more surprised than I that I have been blessed with longevity, but *how* one lives is a choice. I'm reminded of a doctor's visit when I was in my late eighties or early nineties, and he began trying to prepare me for inevitable declining health. He said I'd begin to lose my eyesight and my hearing and that those things are just part of the usual process of growing older. But I thought—and this was before the book, the movie, and all of the extra attention—that I'd lived a wonderful life. And I decided then that while time would do what time does, I would focus on living.

These past few years have been full of wonder. How could I have imagined that from ages 97 to 101, I would be awarded the Presidential Medal of Freedom (with a kiss from my favorite president); appear onstage at the Oscars; receive thirteen honorary doctorate degrees, including one from the University of Johannesburg in South Africa; have four major buildings named in my honor, including a second NASA facility; and need extra storage to house the multitude of plaques, framed certificates, and boxes of mail that would come from all over the world? Many are letters from young people, particularly girls and young women on nontraditional paths, who say my story helped them find the strength to keep going when it seemed everyone doubted them, even at times themselves. Most recently, both Congress and the current US president have approved me to be among the upcoming recipients of the distinguished Congressional Gold Medal. I'm so excited that some of the other African American women trailblazers at NASA—Dorothy Vaughan, Mary Jackson, and Dr. Christine Darden, a longtime family friend—also will receive the medal. Christine is another dynamic and gifted soul, who joined NASA in 1967 as a human computer. By then, racially segregated facilities, including the West Computing Unit, had been abolished at Langley. Women still worked as computers, but the units were no longer segregated. After working eight years as a computer, Christine was assigned to the engineering section and became one of few female aerospace engineers of any race at Langley. She was assigned to write a computer program for the sonic boom as her first task, and ever since she has done groundbreaking research and authored numerous publications on that subject. It will be such an honor to receive the award with her, Mary, and Dot. A fifth and much-deserved group medal also will be awarded on behalf of all women who worked as mathematicians and human computers at NASA.

I can only marvel at these awards. If I've done anything in my life to deserve any of this, it is because I had great parents who taught me simple but powerful lessons that sustained me in the most challenging times. And I was blessed to receive the insight of many others along the way, some who guided me and others who walked the tough terrain at my side.

It's been a remarkable journey, but rarely a lonely one.

chapter 1

NOBODY ELSE IS BETTER THAN YOU

The serious look in Daddy's eyes told me his words were important:

"You're as good as anyone in this town," he said, peering down at my curious little face. "But you're no better."

I was probably in elementary school the first time I heard Daddy say that, and I'd most likely asked him one of my persistent "why" questions, namely why my white playmates, whose fathers worked on our farm, were allowed to call him by his first name. That always irked me because my parents had taught my two older brothers, sister, and me to address all adults as "Mr." or "Mrs." before their last name, as a show of respect. Daddy shrugged off the question, though, as just the way things were. What mattered, he stressed, was what I believed about myself: I was equal to anyone, no matter what the laws or traditions said.

I would hear some version of those words from Daddy repeatedly throughout my childhood, and they helped to shape me to the core. So even as I grew up and was told by law that I had to sit in the back of buses, climb to isolated theater balconies, and use colored water fountains and bathrooms because of my race, I chose to believe Daddy. I was just as good as anyone else, but no better.

Daddy always carried himself like he believed it, too. He was a tall, lean man who walked with a confidence that commanded respect. He wore Stetson hats, which he always tipped at ladies in the community. As a child, I always told my friends that my daddy was the "tallest, handsomest man in the world." He had inherited his family's farm in a small, country town about seven miles outside of White Sulphur Springs, West Virginia. His father, who was of Native American descent, had left the farm to his children when he died. But Daddy's four brothers and three sisters all moved away and left the farm in his care. The farm provided my father's family the kind of independence that was rare for a colored family of that day. Such economic independence and good looks also made Joshua McKinley Coleman a desirable young bachelor in the days before he met my mother.

My mother, then Joylette Roberta Lowe, had been born in Ruffin, North Carolina, but moved to Danville, Virginia, as a child with her family. Her father was a minister, but he died when she was very young. Afterward her mother, who was a talented seamstress, somehow managed to raise four children, including two additional sons and another daughter, on her own. Both girls eventually became teachers. My mother attended Ingleside Seminary in Burkeville, Virginia, which was built in 1892 to educate colored girls. The school had been financed by northern churches and in 1894 was certified as a teacher training institute to respond to the growing demand for teachers to work in schools springing up across southern Virginia for colored children. Girls as young as sixteen years old were graduating from the Institute as teachers.

My mother was eighteen when she started teaching in Virginia. At some point she had a diphtheria vaccination that became infected and left her sickly and with a large, unappealing scar on her arm. Her family figured the pure air from the West Virginia mountains and its freshwater mineral springs would help her recover, and so the teenager moved temporarily with a cousin to White Sulphur Springs.

Located nearly two thousand feet high in the Allegheny Mountains, White Sulphur Springs offered a cooler climate in the summer months than the lower regions, and its natural spring waters were believed to have healing powers. You could smell the minerals when you got close enough to the springs—at least that's what outsiders said. The smell was part of the natural air I breathed, so close to the springs or not, I never noticed a difference. As early as 1778, visitors began flocking to the city to sip or bathe in the springs. A palatial hotel and resort, called the Greenbrier, was developed around the springs, offering premier golf courses, multiple fine dining restaurants, mountain-view pools, and more. The Greenbrier became one of the world's most popular vacation getaways for the wealthy, and with extensive renovations and expansion over the years, it remains so today. Covering thousands of acres—eleven thousand at present—the resort has counted among its guests twenty-seven US presidents, royalty, diplomats, business tycoons, and other high-society family members. For most of its existence, the resort was owned by the Chesapeake and Ohio (C&O) Railway Co., whose passenger trains carried elite private coaches that delivered Greenbrier guests right onto the property. The property also includes ninety-six guest and estate homes, and in more recent years has been used as a training facility for NFL teams, as a professional tennis training center, and even as a concert venue. Locals lived in the town that sprang up outside the resort, and many worked there as servants. But they could not attend any of the events. Daddy later would work at the Greenbrier as a bellman, and he also earned money transporting the resort's monied, white guests around in a buggy pulled by his beautiful horses. So it was at the Greenbrier that the pretty, petite teenager who had come to White Sulphur Springs to heal caught the eye of the area's most eligible bachelor.

She was soft-spoken and tiny, standing just five feet, two inches tall with an eighteen-inch waist, while his gregarious personality matched his tall stature. Their names even matched—Josh and Joy—and they became a couple, courting for a few years before marrying in 1909. She was twenty-two, and he was twenty-seven. She moved with him to the farm, and they lived in a log-cabin-style house in an area known as Dutch Run. My older brother, Horace, was born first in 1912, followed by three additional children: my sister, Margaret (close family members called her "Sister"), in 1914; Charlie in 1916; and me in 1918. I was too young to remember much about our farm life firsthand, but my brothers and sister always looked back on the time with joy. They played outdoors all day, romped through the open fields, and picked huckleberries. Our parents allowed each of them to have a chicken or other farm animal for a "pet," and they helped to care for it.

Daddy raised pigs, cows, chickens, and geese and maintained a large apple orchard. He grew an array of vegetables and fruits, practically everything our family needed to eat off the land and earn a living. Meat was salted down and stored in a shed, or "meat house," for preservation, and some fruits and vegetables were stored in pits underground. To provide minimal refrigeration in those pits, farmers cut large chunks of ice from frozen ponds and covered them with sawdust. Farming was physically demanding work, but Daddy was strong and smarter than anyone I knew.

Mamà (pronounced with emphasis on the second syllable) took care of the chickens, cooked all of our meals, and used a process called "canning" to preserve certain meats, vegetables, and fruits in glass jars for long periods of storage. That was particularly important during the winter months, when ice and snowstorms shut down the roadways and froze everything in place throughout the Allegheny Mountains. At an early age, we children learned how to shell peas, shuck corn, and take the corn off the cob for storage to feed the animals. Our parents bought only a few items at the store, including sugar, coffee, salt, and pepper. But shortly after the United States entered World War I in August 1917, some of those goods, particularly sugar and coffee, suddenly were in short supply.

The bloody conflict had been raging overseas for three years by then, and America's European allies (the United Kingdom, France, and Russia) were practically starving. The war had turned farm workers into soldiers, changed former farmlands into war zones, and greatly disrupted the import of supplies, including food, to the allied forces. When the United States joined the battle against German aggression, President Woodrow Wilson created the US Food Administration to manage the conservation, distribution, and transportation of food during wartime. He appointed then-businessman Herbert Hoover to lead the agency. As the "food czar," Hoover appealed to Americans to reduce their consumption of meat, wheat, fats, and sugar so that there would be more food to send to soldiers. Posters were distributed throughout the country with a targeted message, "Food Will Win the War." Hoover even implemented a system in which Americans abstained from certain foods on designated days—for example, "meatless Mondays" and "wheatless Wednesdays." The success of the effort drastically increased food shipments to Europe and eventually helped to catapult Hoover into the White House as the nation's thirty-first president.

Mamà recalled that the federal government implemented a mandatory draft of men into the US Army during the war under the Selective Service Act of 1917. President Wilson signed the measure into law in May of that year. There were three registrations for the draft, and the third one was on September 12, 1918, for men ages eighteen through forty-five. Daddy, then thirty-five, was required to register in the third group, but he qualified for a hardship provision under the law and was not enlisted. That section of the law exempted married registrants with a dependent spouse or dependent children who would be left with insufficient family income if the breadwinner were to be drafted. I had been born less than a month earlier, and Daddy was the sole breadwinner of our family with four children to feed. I'm sure my parents must have breathed easier after learning that my father would not be forced to go to war. Two months later, relief came for the entire nation when the war ended on November 11, 1918.

During wartime and beyond, the farmers in our community banded together to help one another, regardless of race, especially during the critical harvest season. They worked on one another's farms, shared equipment, and even sat around the table together for meals in one another's homes. In that way our farmers were a different breed of southerners. They understood and respected the hard work, patience, and sacrifice required to earn a living from the land, and they needed one another, all of which often trumped the rules of segregation. When I eventually moved to Virginia as an adult, I always stressed that I was born and raised in "West, by God, Virginia," which we citizens say frequently to emphasize our pride. West Virginia was formed during the Civil War, when it seceded from Virginia to join the Union in its fight against slavery. So to many of the state's African Americans, "West, by God, Virginia" takes on an even more special meaning.

In addition to farming, Daddy bred horses and owned a few of them. The two most beautiful ones were reserved as riding or carriage horses, used to pull our family's surrey, a two-seater buggy with a fringed cover across the top. Somehow, all six of us piled into the vehicle to attend church in town on Sundays. In addition to the carriage horses, Daddy kept two workhorses to help plow the fields and haul timber. He was a skilled logger who cut trees from around the farm and used the workhorses to pull a wagon, loaded with timber, to a nearby sawmill. Despite his limited education, Daddy's mind was quick with numbers—a gift I must have inherited from him. He could add, subtract, and do complicated math problems in his head. He also could look at a tree and instinctively know how many logs he could get from it.

The farm and Daddy's side jobs provided well for our family, but as my brothers and sister reached school age, my parents began to recognize some of the drawbacks of life in the country. The only grade school in the area for colored children was in White Sulphur Springs, and making that seven-mile trip twice daily each day would be disruptive for a busy farmer. So, when I was about two years old, Daddy moved our family into town to make it easier for my older siblings to go to school.

Our new home was much larger and fancier than the old farmhouse, and Daddy had built it himself. The two-story white frame house sat at 30 Church Street, in the middle of a thriving colored community. Everybody called the new place "the Big House," and it definitely lived up to the name. There were four rooms on the ground level, and upstairs featured a bathroom with the first indoor plumbing in the neighborhood. The four bedrooms also were upstairs, one for Sister and me, another for Horace and Charlie, our parents' room, and the guest room. The ceilings were painted sky blue, like Daddy had seen at the Greenbrier. And like the old plantation houses, four columns stood across the front, and broad porches on the top and bottom levels extended the entire width of the house. I enjoyed sitting on the front porch in a rocking chair. Unlike at the farmhouse, we also had a telephone, which was a party line, managed by an operator and shared by several families. Back then, no one had to remember an area code and seven-digit number. Telephone numbers were simple, and I still remember ours: 2–2–8.

Daddy got his hair cut down the street at Crump's Barbershop, which was across the street from Haywood's Restaurant, the only place in town where families from the neighborhood could sit and enjoy a meal. We still attended St. James United Methodist Church, but now we were close enough to walk. The church itself was a plain redbrick building with hard, wooden benches inside and lamplight. Though we were Methodist, the denomination lines blurred a bit out of necessity. Our church and First Baptist Church, which also was just down the street from our house, each had a part-time pastor, and both preached just two Sundays a month. So to attend church every Sunday, most people in the community went to St. James on the even Sundays and to the Baptist church on the odd ones. In the summer, Vacation Bible School was combined, as well as the congregations' summer picnic.

We had many good neighbors, which included for a time Daddy's stepmother, whom we all called Granny. Daddy's mother had died when he was nine years old, and so we never knew much about her. Granny lived down the street, next to Crump's Barbershop, and she made the most beautiful pancakes. My brothers, sister, and I often walked down the street and told her we were hungry so she would make us a batch. In her backyard Daddy kept a small vegetable garden that kept our tables filled with fresh, wholesome food all year long. Though not a blood relative, Granny was our closest connection to Daddy's side of the family. I never knew Daddy's father or his grandfather, but my family would learn later that Daddy's grandfather was a full-blooded Native American. We haven't yet linked our roots to a particular tribe, but the history of Greenbrier County, where White Sulphur Springs is located, can be traced to the Shawnee and Cherokee peoples. Even the origin of city's popular springs is tied to Native American folklore. The Shawnee tribe originally inhabited the forest near the springs, and according to legend, two young lovers slipped away from their elders to the valley to be alone one day. When their chief found them, he became enraged and shot two arrows. One killed the boy, and the other barely missed the girl but struck the ground. The springs supposedly began flowing from the pierced earth. The legend says the girl's lover will be restored to life when the last drop of water is drunk from the springs.

On Mamà's side of the family, her grandmother had been owned during slavery by the white man with whom she bore four boys and two girls. Though it is difficult to imagine today, Mamà said she and her siblings thought well of the former slavemaster, who also was their grandfather. They called him "Bruh" Abe. After slavery ended, Mamà's grandfather quietly kept his colored family together on one side of town, apart from his white family on the other. He also paid for his colored children to attend Hampton Institute, which had been founded in 1868 to educate newly freed slaves. Despite the fond feelings Mamà held for her grandfather, she seemed a bit ashamed of her lineage. She never uttered a word about her grandparents until many years later, when I began digging for answers with my questions. I can only imagine the tortuous conflict that she and other direct descendants of slaveowners must have felt, knowing that their fathers and grandfathers, who afforded them a life typically better than other African descendants, also perpetuated the degradation of slavery.

I loved our Church Street community, but I tagged along with Daddy every chance I got to return to the country. He maintained the farm and often brought bushels of apples from his orchard and sat them on the back porch of the Big House. He enjoyed the woods, and we often found ourselves there, just wandering or picking berries for one of Mamà's delicious pies. Daddy knew the names of every creeping, crawling thing in those bushes. On one of our excursions, when I was about four years old, I felt something warm and slippery underneath my feet and asked Daddy what it was. Before I could bend over to explore, I heard Daddy shout, "Be still! Don't move!"

The next thing I heard was a loud rifle blast. Daddy had shot a snake from right under my feet. "A copperhead," he said, picking up the remains with a stick. As I stood there, still in shock, he pointed out the markings that distinguished this one from other snakes. "They're poisonous," he said, sending the first shiver of fear throughout my body. Daddy had saved my life. I'd always been a "Daddy's girl," but that moment solidified our bond. He would forever walk even taller in my eyes.

I learned quickly, though, that I had to share him, and not just with my family. The entire community, colored and white, seemed to trust Josh Coleman. He knew practically everyone in town, in part because he had multiple side jobs. He began working as a bellman at the Greenbrier when we moved to town, and he at various times cleaned the local electric company, the bank, the Episcopal church, and the library. Also, he kept the vacation homes of some rich, white families who came to White Sulphur Springs in summer to escape the heat and humidity of their hometowns. Daddy kept their keys to come and go, as he needed to maintain their properties while they were away and to prepare for their return in spring and summer. Daddy mostly wore coveralls when he cleaned or worked in the garden, but for church on Sundays and business in town, his preferred attire was a suit and tie with high-top leather shoes. He may have been just a good, reliable servant to the white people who hired him, but Daddy knew his own worth. And he built a cocoon around his children, protecting us from the knowledge that in the outside world, particularly in places such as Mississippi, Louisiana, and Virginia, a colored man could be snatched from his family, dragged before a crowd of bloodthirsty whites, and tortured to death just for carrying himself with a bit too much confidence.

Daddy was well known throughout the community for another reason: he had the gift of healing. People had noticed his special way with horses and began bringing their own troubled animals to him. He could calm even the most agitated horse and cure obscure maladies in them. He was a self-taught veterinarian who treated everyone's animals at no cost. Neighbors and friends began trusting him for their healing, too. I suspect Daddy's special gift and the healing remedies he used were linked to his Native American heritage and traditions. This was an era when there were no doctor's offices on every corner, particularly in neighborhoods where colored people lived, and so people often relied on home remedies handed down for generations. Or they turned to healers like Daddy. The word around town was, if you had something wrong, go to Josh Coleman and he would take care of it. That's what our neighbors from across the street, Robert and Ruth Core, did one day with Mrs. Core's nephew. The boy was visiting from out of town and needed help to stop stuttering, they explained. Daddy spent a few moments talking calmly to the boy, touched his throat, and then pronounced that he would be fine. The boy never stuttered again, according to the Cores, who shared the story with their twin daughters, Annette and Janette. The couple also told their daughters about the time Daddy healed an unusual growth on Mrs. Core's forefinger. The growth looked like a large wart, but it wouldn't go away. Mrs. Core marched across the street one day and explained the situation to Daddy. He quietly took her hand, rubbed his fingers gently across the wart several times, and mumbled something. "Go on home now," he told her. "It will be okay in a couple of days." Sure enough, Mrs. Core told her daughters, the wart disappeared, just as Daddy had said.

Those kinds of stories added a mythical lore to a man who was already bigger than life to me. But I would witness Daddy's "healing hands" myself as an adult when on different occasions he removed a cluster of warts from my baby's eye and my oldest child's hands. Removing warts was his specialty, and his remedy was connected to the *Farmer's Almanac*, which he read faithfully, and based on phases of the moon. Apparently the moon was right for such healing at only a certain time of the month. During that time, White Sulphur Springs residents, colored and white, would make their way to our house, form a line outside the door, and wait for Daddy's touch.

As a child, I was just a little girl in awe of her daddy, and I beamed when friends and family told me I was just like him. I heard that often when it came to numbers. It was well known that I loved counting everything I saw, and I always pushed myself to go higher and higher. Math just made sense to me, and I caught on easily, even before I started school. I loved sitting with my older siblings when Mamà helped them with their homework. Sometimes I'd even come up with the answers before Charlie, who was two years older. One day, when I was about four years old, I slipped away from home. Mamà came outside to look for me, but I was not in the yard. She knew exactly where she'd find me, and there I was, sitting inside the school with my oldest brother.

I vividly remember our neighborhood school, which was initially a two-room building called White Sulphur Grade School. But while I was a student there, the school was rebuilt into a white, wooden, three-room building and renamed the Mary McLeod Bethune Grade School, in honor of the legendary educator and activist who dedicated her life to the advancement of our people. Born to former slaves, Mrs. Bethune was blessed to receive formal schooling and believed education was key to racial advancement. She began teaching in the 1890s and in 1904 established a private school, the Daytona Educational and Industrial Training School for Negro Girls. The school later merged with another institution to form what is today Bethune-Cookman University in Daytona, Florida. Mrs. Bethune became a powerful advocate for civil and women's rights and later would establish the National Council of Negro Women. Her influence rippled throughout the nation, reaching even a West Virginia mountainside, where a tiny schoolhouse, named in her honor, dedicated itself to educating Negro children. Charlie's classroom held a total of about twenty students in first through fourth grades. Horace and Margaret were in the other classroom, with about the same number of fifth through seventh graders.

One of the teachers was named Mrs. Rosa Leftwich, who taught the lower grades. She decided to visit Mamà at our home one day after I slipped away to join my older siblings at school. I was playing nearby as the two women talked, and Mrs. Leftwich began spelling her words to keep me from understanding what she was saying. Mamà gestured frantically with her hands to let the teacher know that was unnecessary. Then, eager to show the teacher what I could do, I burst into the conversation.

"You don't have to spell around me because I can spell," I said cheerfully. "Want to hear me read?"

I had just broken one of the cardinal rules of home training: stay out of adult conversations. But I was so proud of my reading abilities and so eager to show off what I knew that I couldn't help myself. Mamà had taught me to read. My never-ending questions always exhausted her, and to keep my curious mind occupied, she had taught me the alphabet, their sounds, and how the sounds came together to make words. It was like a game to me, and I loved figuring out how to spell everything I saw almost as much as I loved counting. That summer, Mrs. Leftwich started a private kindergarten class in her home with me and a few other children, and I officially started school at four years old. When the Bethune school opened in the fall, I had just turned five and was starting first grade. But it didn't take long for the teacher to discover that I needed more challenging work, and in the middle of the year, I was skipped to the second grade. That wouldn't be the last time I skipped a grade. In the beginning of my fifth-grade year, the school was rebuilt, and a third teacher, Miss Lottie Leftwich, was added. To even the class load, the principal, Mr. Charles S. Arter, visited my class one day and asked if anyone in the fifth grade wanted to move to the sixth grade. Mr. Arter was living in our home as a boarder at the time. There were either no or very few hotels or apartments available to young colored teachers, particularly in small communities, so it was common throughout the South for them to stay with other colored families who had an extra room. As usual, I hopped right up. I always wanted to learn more, and I figured moving ahead to the sixth grade would give me that opportunity. My teacher agreed that I could do it, and just like that, I skipped fifth grade and became a sixth grader.

By the time I entered sixth grade, I had skipped a few grades and was much younger than my classmates. I never thought of myself as advanced, no matter how much of a fuss other people made of me at church and in the community. I always wanted my classmates to know what I knew and was eager to share information. I don't recall ever being scared or intimidated that they were older, and I never doubted that I could keep up. I may have been younger than everybody else, but even then, I knew for sure that I was just as good.

chapter 2

EDUCATION MATTERS

My parents never wavered on their commitment to education, even when it required great personal sacrifice. They gave up Daddy's beloved farm in the country and moved to White Sulphur Springs so that my brothers, sister, and I could attend the only grade school in the area for colored children. Then, as we grew older, they faced another education decision and another huge sacrifice. There was no high school for colored children in White Sulphur Springs or even nearby. Most of the students had to end their formal education at Bethune Grade School after seventh grade. But not the Coleman children. Our parents were determined to do all within their power to give us the best possible education. Maybe we could become teachers, they imagined, because that was as high as they could see. The path to that dream was education. Education was their hope.

So in the mid-1920s, my parents knew their prayers had been answered when they heard about a school, the West Virginia Collegiate Institute, that was offering colored children the chance to get a high school education and a college degree. Sister, a star student, was finishing grade school, so Daddy and Mamà arranged to send her more than 122 miles from home to attend the institute.

Born in the late 1800s, my parents were just a generation removed from slavery, a time when it was illegal for our people to learn how to read, and breaking the rules for a slave could result in a severe beating or even death. Dred Scott wasn't a name in the history books to them. Their parents were alive when Scott, a Virginia slave, had the courage to challenge the laws of slavery for his freedom in a case that made its way to the US Supreme Court. The devastating 1857 ruling, known as the "Dred Scott decision," said that people who had been brought to this country from Africa were not considered American citizens and had no rights under the Constitution, whether they were enslaved or free. The decision added further insult, saying the nation's Founding Fathers believed colored people to be "a subordinate and inferior class of beings" and that white people were "the dominant race." This view remained prevalent among white southerners long after the Civil War and the abolition of slavery and was often used to justify segregation and the mistreatment of colored people.

Nevertheless, my people hung onto their own self-worth and kept pushing to be educated. As early as 1837, a benevolent white philanthropist established the Institute for Colored Youth (now Cheney University) in Cheney, Pennsylvania, the first institution of higher learning for colored students in the country. But most Historically Black Colleges and Universities (HBCUs) were created after the Civil War, by black and white religious organizations and the federal government's Land-Grant College Acts of 1862 and 1890. Those federal laws are perhaps better known as the Morrill Acts, named in honor of Justin Morrill, the Vermont congressman who authored them. They were designed to help states finance colleges that would offer practical training in needed fields, such as agriculture, mechanical arts (engineering), and military science. At the time, colleges tended to focus more on classical studies. The first legislation, in 1862, granted each state at least ninety thousand acres of government land, which were to be sold to generate money for the colleges. However, the measure excluded the Confederate states, which had seceded when the legislation passed during the Civil War. When the war ended, the first Act was amended to include the states that had made up the Confederacy. A second Morrill Act was approved in 1890 to secure the future of those land-grant schools by providing an annual appropriation and to assure that colored people had equal access to the schools. The law required the land-grant schools to show that race was not among their admissions criteria or to establish separate colleges for colored students. A total of sixty-nine land-grant institutions, including seventeen black colleges, were created from both legislations.

West Virginia Collegiate Institute was one of those schools. It was established in 1891 under the second Morrill Act as the West Virginia Colored Institute and initially offered the area's colored students a high school education with an emphasis on vocational and teacher training. The location of the school in the community of Institute, just outside Charleston, was heavily influenced by educator Booker T. Washington, who had spent his formative years in the Kanawha Valley of West Virginia. Washington, who had been born into slavery in Virginia and moved to West Virginia with his family, was by then the founding president of the Tuskegee Institute in Alabama and one of the most influential Negro leaders at the time. He lectured often on the West Virginia campus and even gave the school's first commencement address. Washington believed strongly in the mission of vocational education—so much so that his views, urging our people to focus more on the practical skills that would lead to economic independence, rather than rushing integration, put him at odds with Negro intellectuals of the day.

In 1915, the school was renamed the West Virginia Collegiate Institute to reflect its new college degree status. Four years later, a confident young intellectual named John W. Davis was hired to take the school to the next level. Davis had graduated from the prestigious Morehouse College in 1911, studied at the University of Chicago, served on the faculty at Morehouse, and was executive secretary of the Twelfth Street Branch of the Young Men's Christian Association in the nation's capital at the time of his appointment. When he arrived at the Institute, less than half of the 24 faculty members had college degrees, and just 9 percent of the 297 students were enrolled in college courses. He immediately began recruiting highly educated professionals and strengthening the curriculum. Among the recruits was Dr. Carter G. Woodson, who had earned a PhD in history from Harvard University in 1912. Dr. Woodson was the second Negro to earn a doctorate from Harvard (after W. E. B. DuBois, who became a prominent historian and scholar). Dr. Woodson served as academic dean at the institute for two years, from 1920 to 1922, and dedicated the rest of his life to teaching and helping others understand the history of our people. He would go on to establish Negro History Week, which ultimately evolved into the celebration known today as African American History Month. He would return to the college as a speaker during my time there.

It must have been difficult for my parents, providing for Sister so far away, while maintaining our family at home. But somehow they made it work for the next two years. Then Charlie finished grade school, and once again my parents made a life-altering decision for the sake of their children's education. They agreed to leave behind the comfortable life we'd known in White Sulphur Springs and move our entire family to the town of Institute so that all of their children could attend the school. At about this time, Daddy also sold the farm. For a man whose life and history had been so connected to that land, giving it up must have been a tough decision. But there was something else just as important to him: education. When Daddy was in school, colored children could go only as high as the sixth grade. He wanted to be sure that his children got the education he had been denied. Our family didn't have a home in Institute, nor did Daddy have a waiting job. But my parents had what mattered most: lots of determination, dreams, and hope. With an education, just maybe their children's lives wouldn't have to be limited to cleaning, cooking, and serving the rich white guests at the Greenbrier. Maybe we could have something better. So in September 1928, the six of us piled our belongings into a truck with a driver Daddy had hired for the trip, climbed inside, and waved good-bye to the Big House and our old life.

Once our family arrived in Institute, we stayed for a while with Mamà's cousin, Nannie, and then rented a house from Nannie's husband. Mamà enrolled Horace and Charlie with Margaret in high school, and I entered the primary school. Both schools were on the Institute's campus, across the yard from the buildings where the college students attended classes. A critical part of my parents' plan did not go as expected, though. No matter how hard Daddy tried, he could not find work. He even met with Dr. Davis and asked if there was available work on campus, but the answer was an unfortunate no. This must have been quite a surprise and disappointment to a man who always had been able to provide for his family by working multiple jobs. But this town on the Kanawha River was not White Sulphur Springs, where Daddy had lots of connections and everyone knew about his multiple talents and stellar reputation. Still, giving up was not an option. Our parents just came up with another plan. Daddy returned to White Sulphur Springs and the Big House alone and began taking on extra work so he could maintain two households.

My parents' actions in those years spoke to me louder than any words they ever said about how much education mattered to them. They were sacrificing mightily, and that set the expectations high for my siblings and me. How could we do anything but work hard in school when our parents were working so hard to provide us such an opportunity to be educated? My brothers and sister also worked hard outside school by getting odd jobs to help with our tuition and expenses. Horace and Charlie became newspaper and milk delivery boys, and they worked on campus. Sister assisted Mamà, who had begun ironing white people's clothes in our home, a common way for colored women of that era to earn money. Mamà was such a talented seamstress that she sometimes made us replicas of the dresses she and Sister saw in their pressing piles. At just ten years old, I took on more chores around the house to do my part. I had never been separated from Daddy for such long periods of time, and missing him was the most difficult challenge of our new life. I wrote him letters to share the details of our days, and I looked forward to summer and long holidays, when the five of us returned home to White Sulphur Springs to be with him. This would become our routine for the next decade or so, until all four of us children had completed high school and four years of college.

Eventually we all adjusted to our new reality. I loved being on the Institute's campus, which was like its own community. But to a country girl who had never traveled much farther than White Sulphur Springs, this was a different kind of place. One of the first things I noticed was that every class had a different room and a different teacher. I had never before seen so many educated colored men and women, as well as students pursuing high school diplomas and college degrees, all in one place. There were other students like me, who were carrying the dreams of parents whose own education had been stunted, and there were children of privilege whose parents were middle-class teachers, scholars, and business owners. We were on this journey together. Many of our instructors had studied at colleges in the North, and they even sounded smart with their big words and perfect diction. The men wore suits and ties, and the women wore dresses and heels. They believed we all were part of the "Talented Tenth," the exceptional, highly educated class of Negro leaders who would lift our race, as described by W. E. B. DuBois in a much-publicized 1903 essay. This gave our instructors a sense of mission and purpose. They got to know us, but they also challenged us and demanded excellence.

One of those teachers was Mrs. Rose Evans, who was hired to teach my eighth-grade class when the regular teacher became ill at the beginning of the school year. Mrs. Evans was married to Professor James C. Evans, head of the college's Mathematics Department. Professor Evans was a superintelligent man who was born in Tennessee and graduated from Roger Williams University there. He also received bachelor's and master's degrees in electrical engineering from the Massachusetts Institute of Technology. Mrs. Evans was just as smart and had a teaching degree, but in those days, married women were expected to be homemakers and to focus on their families. Many schools, including the Institute, even had a clause in their teaching contracts that allowed a female teacher to be fired if she got married. Perhaps out of necessity, the Institute had relaxed the rule that year, enabling Mrs. Evans to teach my eighth-grade class. My good friend and classmate Constance Davis, nicknamed Dit, and I were both excellent students, and we were always eager to help our teacher with after-school chores. That must have caught Mrs. Evans's eye, and she began inviting the two of us to her home for tea, homemade cookies, or cakes. The couple lived a short walk from campus, and Dit and I visited often. Mr. and Mrs. Evans were both great motivators, and they made us feel special. They talked to us about the importance of doing well in school. As colored students, we represented more than ourselves, they said. They reminded us that we carried on our shoulders the hopes of our entire race. Our successes would help to dispel the lies of whites that our people were inferior. When we rose, so did our people.

"You have to be better today than you were yesterday," Mrs. Evans often said. Then she challenged us: "What do you know today that you didn't know yesterday?" To two curious girls, who loved a good challenge, Mrs. Evans's words became like a game, and we hungrily sought new information and current events so we were ready to respond when she popped the question.

It wasn't difficult to learn something new on the campus of the Institute, which was full of interesting history, including the story of how the state acquired the land to build the school in the first place. It dates back to a rich, white slaveowner and the enslaved woman with whom he shared his life, riches, and many children. The slaveowner was Samuel I. Cabell, who was connected to the powerful Cabell family that in Virginia includes military generals, politicians, judges, and even a former governor. But Sam Cabell left a different legacy. He fathered at least thirteen children—some accounts say the number is fifteen—with one of his slaves, Mary Barnes. Though Sam owned his woman and children all of his life, he is said to have devoted his life to them. He lived with them and other slaves on a plantation he developed along the Kanawha River, just outside Charleston. Between 1851 and 1863, he went to great lengths to protect Mary and the children legally, writing several different wills that granted them freedom upon his death and bequeathed to them all of his riches. But the same slaveowner penned an angry diatribe in his final will in 1863 against the slaves who had fled his plantation or were taken away by federal troops. He rescinded an earlier will that had laid out plans for the freedom of his nonfamilial slaves, and he declared that the runaways should remain slaves for the rest of their lives. Sam was murdered on July 18, 1865, about two months after the Civil War. Though there was much speculation that he was killed by white townspeople who disapproved of his affection for his interracial family, the reason for his murder was never determined. Several men, who were known to be Union sympathizers, were arrested in the killing, but they claimed self-defense and were acquitted.

Sam's wills were later determined to be valid, and Mary was successful in her petition to the court to have her last name changed to Cabell. The Kanawha property was divided among Mary and the couple's children. In 1891, when the state created the West Virginia Colored Institute under the Morrill Act, officials initially had trouble finding a location for the school because white landowners wanted no part of it. But when one of Sam's daughters, Marina, heard about plans for the school, she ultimately sold the state the first thirty acres for the project, and the state gradually acquired another fifty acres to build an eighty-acre campus. The Cabell descendants recognized the value of the West Virginia Colored Institute because there was no other school in the entire state or nearby fulfilling such a mission. Sam and Mary had been forced to send their own children as far away as Ohio to be educated. Some of them ultimately returned to their Kanawha River community as doctors, teachers, and other professionals, and their descendants built one of the largest communities of colored people in West Virginia. Sam and Mary Cabell, as well as some of their children, are buried in a family cemetery on the grounds of the Institute.

I loved exploring that campus with my best friend, Dit, who was smart, funny, and down-to-earth despite her upbringing in a well-educated, well-to-do family. In some ways the two of us were from different worlds. She was the daughter of our highly regarded college president, Dr. Davis. He and his wife, Bessie, had two daughters—Constance and Dorothy—nicknamed Dit and Dot, a playful reference to the military's secret telecommunications system (Morse code) and the hard work it took to keep up with two fast-moving girls. The family lived on campus in the president's residence, called East Hall, which had been built in 1893. It was a two-story, white frame, antebellum-style house, and though it was much more elegant, it reminded me a bit of the Big House in White Sulphur Springs.

Under Dr. Davis's leadership, the Institute continued to grow and attain greater status. In 1927, the year after my family arrived, the school became the first of the seventeen original black land-grant schools to be certified by a regional education association, the North Central Association of Colleges and Schools. Dr. Davis and his staff also were very proud that this certification made the Institute the first public college in West Virginia, black or white, to be accredited by the organization.

My first year at West Virginia Collegiate Institute passed quickly, and before I knew it, I finished eighth grade at just ten years old. I was so excited about attending high school, partly because I adored a certain math teacher, Angie Turner. Miss Turner had taught my siblings both algebra and geometry, and those subjects fascinated me. I often sat next to Sister at the kitchen table and watched eagerly each night as she pored over her homework, figuring out equations and calculating angles. Of course, I asked question after question, and she patiently answered each one. Sister was just as smart as everyone said I was, but she was very shy and quiet. She had a condition we called "lazy eye" because one of her eyes did not move properly, which may have taken away some of her confidence. Sister talked to me endlessly about Miss Turner, and because Sister loved her as a teacher, I knew I would, too. But to my great disappointment, Miss Turner left the Institute to pursue an advanced degree just as I was entering high school. Many of the Institute's teachers, from the primary school instructors to the college professors, had their master's or doctorate degrees. So as bad as I felt that I didn't get to have Miss Turner as a teacher for algebra and geometry, I understood our teachers' desire to improve themselves continually. They had been taught—and they taught us—that to have a chance in this world, we as colored people had to be twice as good as our white counterparts. Miss Turner earned her master's degree in chemistry from Cornell University in 1931. By my senior year in high school, she had returned to the Institute, and I had the pleasure of being a student in her chemistry class. When we went off to college, Miss Turner, just thirteen years older than I, allowed her former students to call her Angie.That made me feel like a grown-up.

Angie Turner is a perfect example of the caliber of teachers I was so fortunate to encounter on my journey. She eventually was offered a teaching position on the college side of the Institute, where she continued mentoring students. In 1946, she got married, becoming Angie Turner King. She later earned a doctorate from the University of Pittsburgh and would spend her entire career at the Institute and the rest of her life in that town.

I continued to excel throughout high school, including in my nonacademic courses, such as home economics, which taught girls how to cook, sew, and manage a household. Sister and I had learned most of those things already at home in the kitchen with Mamà, who sewed most of our clothes and baked delicious cobblers and cakes. The prospect of learning something new always excited me. I learned how to read music and play the piano during music classes at school. I learned how to play tennis by tagging along with my brothers and bombarding them with questions. They also taught me how to ride a motorcycle and drive a car later in life, at a time when most women still did not do such things. I suspect that too often the fear of failure keeps people from trying to learn new things, and they miss out on wonderful experiences.

High school was full of excitement for me, but there was sadness, too. On February 24, 1931, Dit lost her mother to liver cancer, and our campus lost our First Lady. It was such a sad time. Mrs. Bessie Davis was an intelligent, dignified woman who had been born into a wealthy Negro family in Atlanta. I would learn later that her father, Henry A. Rucker, had been born into slavery in Washington, Georgia, but moved to Atlanta with his parents and siblings after the Civil War. He became educated and opened some businesses, including a barbershop that was frequented by white politicians. In 1897 he was named the head of revenue collection in Georgia, a powerful position for a Negro during the Reconstruction era. Mr. Rucker held that position until 1911, becoming one of the most powerful Negro Republicans in Atlanta. At the time of his death in 1924, he was widely considered one of the wealthiest Negroes in the country. When the daughter of this powerful politician died seven years later, at just forty years old, the Rucker family held the funeral in the family home in Atlanta. Many of the faculty and staff from the Institute traveled there to support Dr. Davis and his family. I wasn't totally aware of Mrs. Davis's prominence at the time, though. To me, she was simply my best friend's mom, and at just twelve years old, I hardly knew what to do or say to help Dit cope with such a huge loss. But happier times did come when Dr. Davis remarried in September 1932. His new wife, Ethel McGhee, was an accomplished woman who had been Dean of Women at Spelman College, the sister school to Morehouse in Atlanta. The students loved her as well, and when the couple later had another daughter, named Caroline, the students nicknamed her "Dash." So that's how the Davis girls became Dit, Dot, and Dash.

Through it all, I never lost sight of my parents' determination to keep my siblings and me at the Institute, especially during the Great Depression. When the stock market crashed on October 29, 1929, the nation's economy spiraled, causing banks to fail, businesses to close, farms to foreclose, and millions of people to lose their jobs. Thousands of families across the nation relied on soup kitchens and the generosity of strangers for food to survive. The suffering was even more acute in West Virginia with the collapse of its coal mines, the state's main industry. In some parts of the state, the unemployment rate was as high as 80 percent, and people were struggling mightily. Like everyone else, my family had to cut back on the food items we usually bought at the store and stretch our meals as much as possible, but Daddy still managed to keep us all fed and clothed. I don't think I realized then just how fortunate we were, even in those meager times. Fortunately for Daddy and the other servants at the Greenbrier, the resort still bustled with rich guests entertaining (or perhaps consoling) themselves on the golf courses and in the fine restaurants, pools, and spas. They still wanted their bags carried and unpacked, their children watched, and their carriage waiting, so there was work. During the summers, Horace and Charlie joined Daddy at the Greenbrier as bellboys, and Sister worked in the valet shop, taking care of the guests' personal services, including ironing their clothes. When I grew older, I joined her there. In the meantime, I helped out by getting a job on campus in President Davis's office during the school year.

I graduated from high school in the spring of 1933 and started college the following fall at just fifteen years old. By then, the school's name had been changed again, to West Virginia State College, and its enrollment had grown to about a thousand students. My family lived in a house just across the street from campus, and it was ideal for me. I loved being around smart people. Of all the Coleman kids, I was the busybody. I got to know practically every student at the university by name, in part because I met people along the way as I roller-skated from one end of the campus to the next. When one of the first highways was being built through Institute, we students skated right on top of it. On Saturdays, small groups of us enjoyed hiking through the mountains. We just made our own paths through those woods. In 1934, I would even follow Dit into the sisterhood of Alpha Kappa Alpha Sorority, Inc. My sister and brothers had no interest in joining a sorority or fraternity. But I loved being part of such an esteemed group. The AKAs, as we were called, were popular on campus, but beyond that, I found a lifelong tribe of like-minded ladies who are committed to promoting education and serving our community.

Meanwhile, West Virginia State was making a name for itself and becoming a focal point of intellectual and cultural activity for Negroes in the area. All college students were required to attend "Sunday Night Chapel," and it seemed all of the famous Negro scholars, activists, and entertainers made their way there to visit at one time or another. Negro colleges and universities were the high-end, academic equivalent of the "Chitlin Circuit," presenting our community's best and brightest to malleable minds. I recall such bright lights as Paul Robeson, Marian Anderson, George Washington Carver, Benjamin Elijah Mays, Mordecai Johnson, and W. E. B. DuBois. Many of them stayed at the Davis home, and because I was in and out of there all of the time with Dit, I got to meet them personally. At the president's house they were just like everybody else, and I must admit that I was too naive even to be starstruck. Some of the speakers were lesser known, but they were passionate in their crusade against one of the biggest issues of our time: lynching. It was shocking to hear that our people were being shot, hanged, and tortured in public venues all over this country, particularly in the South, for minor infractions or none at all, as entertainment for the participating white mobs.

One of the most disturbing cases occurred in late December 1933, when seventeen-year-old Cordie Cheek was lynched in Maury County, Tennessee. The teenager had been falsely accused a month earlier of raping a white girl in Maury County, but a grand jury refused to indict him for lack of evidence. After being released from a Nashville jail, he went to stay with an aunt and uncle in Nashville, on the edge of the Fisk University campus. Three white men, including the county magistrate, showed up there, took him from the home, and transported him back to Maury County, where an excited white mob was waiting. The boy was blindfolded, castrated, and hanged from a cedar tree. The crowd then passed around their pistols and celebrated. Historian John Hope Franklin, who was a student at Fisk at the time, recalled those horrible days decades later in his 2005 autobiography: "Those of us who had remained in Nashville over the Christmas holidays were obsessed with discussing the Cordie Cheek lynching. Indeed, the entire remainder of our junior year was shadowed by this tragic event. There were investigations, interviews, and other actions. The conclusion that many of us reached was that if it could happen to Cordie Cheek, who had been seized within three blocks of the Fisk Chapel, it could happen to any of us."

Word about such incidents traveled quickly among colored college students on campuses across the country and prompted many of them to get involved with the campaign of the National

Association for the Advancement of Colored People (NAACP) to end racial violence against our people. These were weighty issues that reminded us of our dangerous times. Yet we were still young, idealistic, and hopeful for our futures. Heavy on my mind was choosing a major. As a college freshman, I wasn't required to declare a major right away, and I was unsure which one I would choose, so I didn't enroll in math. My grades were proof that I was just as good in all of my subjects as I was in math, but I had fallen in love with English and French, especially French. I loved the exotic way the language sounded. Soon I would love it even more.

After my freshman year in college, my family returned home to White Sulphur Springs for the summer, as usual, and I joined Sister, working in valet services at the Greenbrier. I was unpacking and ironing a guest's clothes one day when a woman, who had identified herself as Countess Sala, began talking on the phone to her husband in French. My ears perked up. At one point I must have looked up at her in a knowing way. Our eyes met, and I quickly looked down. I hadn't meant to stare. Her conversation stopped. She asked if I understood what she was saying.

"Yes, ma'am," I responded, unsure what would happen next.

She didn't seem upset and began asking me lots of questions—my name, age, where I'd learned to speak French, and more. I felt relieved that she wasn't angry. I answered her questions and told her I was a student at West Virginia State College. She seemed impressed and arranged for me to practice speaking French with a Parisian chef in a restaurant at the resort. It was an amazing opportunity. The chef and I agreed to meet once a week, and I could hardly wait for each session. He corrected my diction, taught me new words and phrases, and most of all, helped me become conversational in French. That summer I added a French accent to Mamà, and that became my name for my mother. The rest of the family picked up on it, too, and it lasted for the rest of her life. By the time I returned to West Virginia State, my French had improved dramatically, and I was certain that French would be my choice for a major. But then the math professors had their say.

I was walking across campus one day the next semester and saw a teacher named Mrs. Lacey, whom I recognized from my visits to Dr. and Mrs. Evans's home. Mrs. Lacey was a math teacher who had been living with the Evans family as a boarder. But shortly after I met Mrs. Lacey, she got married and lost her job at the Institute because of the marriage clause. I was thrilled to see her back on campus. The school had lifted its marriage clause to allow openly married teachers to join the faculty. I would learn later firsthand that some teachers felt a need to hide their marital status from their employers to keep their jobs. Mrs. Lacey seemed genuinely happy to see me, too.

"Katherine, I'm teaching math again," she said. "If you aren't in my math class next semester, I'm coming after you."

We both smiled. "Yes, ma'am," I replied.

It was time to register for the next semester's classes, and I quickly changed my schedule to take her math class. I had been enrolled in the class of another new, young professor, but I withdrew from his class and kept my word to Mrs. Lacey. She was a fantastic teacher. When I signed up the following semester to take the next math class under the new professor whose class I had dropped, he told me exactly what he'd thought of that decision. Apparently the word had gotten around the Mathematics Department that I was a promising math student. "I was so furious," the professor told me, feigning anger. "I thought I had this good math student and you get up and go to Mrs. Lacey's class!"

He was Dr. William Waldron Schieffelin Claytor, and he had just received his PhD in math from the University of Pennsylvania the previous spring of 1933. That made him only the third Negro in the entire country to receive a doctorate degree in math. The previous two doctoral recipients had taught him at Howard University, where he received his bachelor's and master's degrees. In his mid-twenties, Dr. Claytor was well dressed, confident, and very good-looking. All of the girls on campus just swooned over him when we spotted him playing a vicious game of tennis or driving onto campus in his sharp sports car. But Dr. Claytor was, above all else, a fabulous mathematician and teacher. He would stroll into the classroom, reach into his pocket for a piece of chalk, and walk immediately to the blackboard. It always amazed me that he knew exactly where the class left off the day before. He stopped at his desk only to get the textbook out to give our assignment for the next day. I was mesmerized, so I kept signing up for his classes. Sometime after that first year he said to me, "You know, you would make a good research mathematician."

I hadn't heard that term before, so I asked, "What does a research mathematician do?"

He chuckled and replied, "You do research in math."

Before college, I didn't even know colored people could get a job higher than teaching. When Daddy and Mamà would suggest to me that I should be a teacher, I would tell them I wanted to be a *college* teacher, not even sure back then what college was. But it sounded important, so I figured a college teacher must be an extra-big job. But a research mathematician? I had to ask:

"Well, where do you get a job doing that?"

"That will be your problem," Dr. Claytor responded. "But I will have you ready."

And he made sure I was. When I had taken every math course the college offered, he said I still needed a couple more, and he added them to the curriculum. The final class I took from him was one that he created just for me, and it was called Analytic Geometry of Space. I couldn't have known then how prescient that move would be. And it never occurred to me to ask him why I might need such a class. None of my schoolmates was interested enough in math to take the advanced course, but Dr. Claytor didn't give me an option. "You're going to take it," he said simply. I was the only student, but he taught that course as if the entire room were full of people. I was fascinated by this math. Dr. Claytor stretched my mind farther than I ever believed possible, but the thing I loved about math, more than any other subject, was that there was a definite right or wrong answer. It was sometimes hard to figure out, but when you got it, you knew it, and what a wonderful feeling.

Though I didn't realize this at the time, Dr. Claytor must have been doing for me what two of his professors at Howard University had done for him. A young W. W. Schieffelin Claytor entered Howard as an undergraduate in September 1925, after attending public schools in Washington, DC. He'd been born in Norfolk, Virginia, into an educated, middle-class family. His mother, Simsie Thorne Claytor, had graduated from Hampton Normal and Agricultural Institute in Hampton, Virginia, in about 1900. His father, William Oat Claytor, initially worked as a vocational arts teacher at a segregated school in Norfolk, Virginia, before relocating his family to the nation's capital. There the elder Claytor earned a dentistry degree from Howard in 1916 and ultimately opened his own dental practice. The son followed his father's path to Howard in 1925, majored in math, and earned a bachelor's degree in 1929. The same year, Dr. Dudley Weldon Woodard Sr. established Howard's first master's degree program in math, and Schieffelin Claytor, as he was called then, became one of the first students. Dr. Woodard, who had served dual roles at Howard as a math professor and dean of the College of Liberal Arts in the early 1920s, was just returning to the university from a leave of absence in the 1927–28 school year. He had spent the time finishing his doctoral studies to earn a PhD in math from the University of Pennsylvania in 1928. He became the second Negro to earn a doctorate degree in mathematics.

The first to hold that distinction was Dr. Elbert Cox, who earned his PhD from Cornell University in 1925. Ironically, Dr. Cox spent his first four years teaching at West Virginia Collegiate Institute before joining Woodard at Howard, just as the new master's program was beginning. So an eager, young Claytor must have seen a clear route to at least some of his aspirations as he studied under the only two Negro professors in the country with PhDs in math. In 1930 he would receive his master's degree from Howard and follow Woodard's advice and strong recommendation to the University of Pennsylvania. No doubt he proved himself to be a brilliant student, earning free tuition and a stipend with the Harrison Scholarship in Mathematics in his second year and the prestigious Harrison Fellowship in Mathematics in his final year. He completed the doctoral program at Penn in 1933—the third Negro in the country with a PhD in math, after his two Howard professors.

Even as Dr. Claytor accepted the teaching position at West Virginia State the same year, he had a higher ambition, the same ambition that he would spark in me: to become a research mathematician. But a relatively small, isolated college with limited resources, budget, and staff was no match for his aspirations. So as Dr. Claytor was preparing me, he was calculating his own exit.

Dr. Claytor would take a one-year leave of absence from West Virginia State in 1936 to pursue post-doctoral work at the University of Michigan. He was thrilled to join a group of mathematicians who were doing interesting work in his field of interest, topology. A Rosenwald Fellowship helped to finance his research over the next three years, but he would face heartbreaking disappointment and racism. In 1936, for example, he was invited to present his research at the annual conference of the American Mathematical Society in Chapel Hill, North Carolina, but he was not permitted to stay at the hotel with other members because of his race. He stayed with a Negro family in town. His credentials didn't matter. At that time, in that place, his race alone determined his status. The incident so hurt him that he eventually stopped attending the society's conferences.

The next year, Dr. Claytor's research was published in the *Annals of Mathematics*, which some sources credit as the first published math research by a Negro aside from degree theses. His work would be cited by researchers for years to come, and time eventually would prove him to be a mathematics genius. But Dr. Claytor's fellowship funds dried up in 1939. With the help of famed math professor Raymond L. Wilder of the University of Michigan, who was among his white advocates, Dr. Claytor stayed in Ann Arbor and managed to get enough temporary work to support himself a couple more years. Still, he could not find a permanent teaching position at any research institution, even when one opened at the University of Michigan, where he had spent the past few years. Dr. Claytor had no desire to teach at small, black colleges in the South because the demands of teaching a full load of classes left no time for research, his true passion. But he would find, perhaps like his two predecessors, that race limited his options. At one point in 1939, Wilder received good news that the Institute for Advanced Study in Princeton, New Jersey, would find a place for Claytor. But five years earlier, while Claytor was still at West Virginia State, the same organization had been housed at Princeton University, and when Claytor applied to conduct research there under a fellowship, university administrators rejected him. White students might object to the presence of a colored scholar, they reasoned. Though the Institute for Advanced Study had by 1939 moved to an independent location in Princeton, Claytor was still stinging from the earlier rejection. Wilder would express in a letter to a colleague many years later that when he approached his young protégé about the chance to work at the research organization, Claytor "shook his head and replied, 'There's never been a black at Princeton, and I'm not going to be a guinea pig.'" No other opportunities were forthcoming.

Karen Hunger Parshall, a history and mathematics professor at the University of Virginia, later studied Dr. Claytor as a detailed case for a research paper titled "Mathematics and the Politics of Race," published in the *American Mathematics Monthly* in March 2016. In the article she concluded, "Claytor may have had strong and influential supporters within the American mathematical community, but he understood and lived day-to-day under the social forces at work in the immediately pre- and post-war United States, social forces that had yet to allow for the integration of professional meetings much less the faculties of the nation's research universities. . . . Claytor had been trained to do mathematical research, but de jure segregation in the South and de facto segregation elsewhere, left the research universities largely closed to him and made it hard for him even to participate in the activities of his professional societies."

Dr. Claytor joined the US Army in April 1941, just months before the United States entered World War II. He then did short teaching stints at two black colleges in the South and married Mae Belle Pullins, an accomplished professor who earned a PhD in education and established herself as an authority on the issue of standardized testing. Dr. Claytor never returned to West Virginia State. By 1947 he wound his way back to Howard University—just as his previous two math professors had done—and spent the rest of his career there, teaching math. He died with a broken heart at age fifty-nine.

I graduated summa cum laude from West Virginia State in 1937 with degrees in both math and French and the highest grade point average of any previous student in the school's history. Dr. Claytor had left the previous year to pursue his passion. In some ways, we both left there starry-eyed, in search of an answer to the problem he'd dropped in my lap: how to become a research mathematician as a colored person in 1930s America. In time he would draw his own complex conclusions.

For now, though, the problem was mine to solve.

chapter 3

A TIME FOR EVERYTHING

For everything there is a season, a time for every activity under heaven.

—Ecclesiastes 3:1

God has an appointed time for everything. That's what the Bible says in one of my favorite verses. As my college graduation approached, I didn't know the first thing about how to become a research mathematician, so I figured it clearly wasn't that time. I decided to look for a job doing what felt more familiar: teaching.

The nation had not yet pulled itself out of the Great Depression, and jobs were scarce. Even in more prosperous times, Negro teachers had been limited to working in the few schools available for Negro children. But in times of such economic uncertainty, workers everywhere were hanging onto their jobs. With so many people out of work and suffering, the teachers with jobs must have felt fortunate. Plans to pursue advanced degrees or relocate to different areas were being put on hold, and eager college graduates hoping to become teachers were finding few openings.

That was on my mind in the first months of my senior year,

when I realized I had enough credits with the classes I was taking to graduate in December 1936. Many seniors may have jumped at the chance to finish college early, but the more I thought about it, the less it seemed like a good idea. If I graduated early with no job offers, I would have to return home to White Sulphur Springs. That would add another mouth for Daddy to feed. As things were in my last year at West Virginia State, I had a full-tuition scholarship that also covered my room and board. If I stayed on campus an extra semester, it would not cost my parents a thing. So I dropped a class, stayed the next semester to take that same course, and bought myself a little time. Still, when I graduated in May 1937, I ended up returning to White Sulphur Springs with my college degree, every honor I could have received, and no job prospects. June and July passed, and still nothing. Then, on August 4, 1937, I received an urgent telegram that said only this:

"If you can play the piano, the job is yours."

The name on the message wasn't familiar. I immediately called Mr. Evans, the Mathematics Department head at West Virginia State, because Negro schools that needed teachers often reached out first to the area's Negro colleges for recommendations.

"Did you recommend me for a job teaching piano?" I asked him.

He had not. I thought of another professor who may have referred me and called him. Yes, he had made the recommendation, he said. A school in Marion, Virginia, was looking for an elementary teacher who could teach French and piano to students in grades four through six. The professor said he remembered that I had taken piano lessons and thought I would be perfect for the position. With this new information, I called to accept my first teaching job, at Carnegie Elementary School, and began preparing to move 127 miles from home. The opening of school was just weeks away. I was nineteen years old, and soon I would be a brand-new teacher.

Just before I left, Mamà issued this warning: "Remember, you're going to Virginia."

In our minds, Virginia was the real South. So in other words, she meant, "Remember your place and act accordingly."

I responded flippantly, "Well, tell them I'm coming!"

Soon I was aboard a bus and on my way. Just as Mamà had warned, it didn't take long for me to realize that I had crossed the line into Virginia. The bus came to a stop abruptly, and the driver shouted some orders: Negroes to the back of the bus. We had been interspersed with whites on the bus, but as routinely as getting off at the appropriate stop, all of the Negroes stood and moved to the back, as a few white passengers moved closer to the front. I watched in stunned silence and followed suit. Later, when it was time to change buses, the white passengers were allowed to board a new bus, but the driver shouted my way, "All you colored folk, come over here!" I pretended not to hear him until he softened his tone. The buses did not go into the colored side of town, he said. Negro passengers then had to pay a taxi or find another way to get where we were going. Everyone seemed to know the rules, and a few Negro taxi drivers were waiting. I climbed into one of the cars and headed into my future.

Segregation certainly was not new to me, but I was not accustomed to such blatant racist and rude behavior. I refused to let it bother me, though. I just stared out the car window and watched this new town unfold before me. My first stop was the principal's home, where I arranged to rent a room in his neighborhood from a woman named Mrs. Gert Ross, whom everybody called Aunt Gert. She lived in a rather large house that was walking distance, about a mile and a half, from the school in a hilly community known as "Up the Creek." As the name implies, a small creek ran through the neighborhood.

Marion was a picturesque small town in the highlands of southwestern Virginia, near the borders of Tennessee and North Carolina. Named in honor of Francis Marion, an American Revolutionary War hero, the town is the county seat of Smyth County. About five thousand people called Marion home in 1937, and the vast majority of them (at least 96 percent) were white. There were a small number of Negro families, who lived in mostly segregated areas of town, but that wasn't the case everywhere. Charles and Anna Goble and their thirteen children were the only Negro family living on a hill in an established neighborhood about two blocks from Main Street. The younger Goble children, one of whom was in my class, often played in the neighborhood with white children, including the sons and daughters of a couple of local and state politicians. When Anna Goble sat on the porch, the white mothers usually stopped to chat. But this was the South, and segregation was the law. So while Charles, a second-generation barber, could cut white men's hair in a whites-only barber shop on Main Street, he had to cut even his own children's hair at home.

When I arrived, there was still much buzz around town about the new Hungry Mother State Park, which had just opened the previous summer. The park was 1,881 acres of donated land that had been transformed into a stunning playground with picnic areas, cabins, hiking trails, a pool, and a bathhouse. It had been constructed by the Civilian Conservation Corps, a national relief program that was part of President Franklin D. Roosevelt's "New Deal" to provide jobs throughout the country and help jump-start the depressed economy. Negro residents were happy that the park had brought more jobs. Once it was open, they could work there as servants, but their families were not allowed to visit and enjoy the festivities as guests. Another of Marion's local treasures was the three-story, red-brick Lincoln Theatre, with its Mayan temple–inspired Art Deco theme. The movie palace had been built by the town's wealthiest resident, Charles C. Lincoln Sr., and offered new films and vaudeville acts to a community starved of that kind of culture. The theatre opened just months before the crash of the stock market and somehow managed to survive during the Great Depression. Many Negro patrons enjoyed attending the shows, but they had to enter through a back door and climb the stairs to the balcony. Such were the degrading rules of segregation, but the Negro families in town just built their own social lives, centered mostly around their families, friends, and church.

When Aunt Gert took in a boarder, she became our family away from home. She cooked our meals and introduced us to the community. This was my first time living on my own, so I focused all of my energy and attention on preparing for my students. My wardrobe was already set. I'd used some of Mamà's best patterns over the summer to make a few pretty, new dresses for work. I wanted to look like a teacher. Each morning, after putting on my dress and heels, I always dabbed on some lipstick. The girls and boys in my class seemed so enamored by their new teacher. I sometimes caught them gazing up at me and giggling in their cute, childish way. I wouldn't realize until many years later that a big part of their fascination was the lipstick. In this small town, my students were not accustomed to seeing a woman wear lipstick, and in their innocent eyes, painted pink lips made me seem a bit daring. I had no idea. More than anything, I wanted to do a good job for my students. I realized that most of them would not be able to extend their education beyond the Carnegie school.

Carnegie was a two-story, redbrick building perched atop a hill. The main entrance was at the top of a flight of stairs on the second level. There were four classrooms, two on each side, divided by a long hall. The elementary grades were on one side (one through three in one room and four through six in the other), and the upper grades were on the other side (seventh, eighth, and ninth in one room and tenth and eleventh in the other). Each grade had an average of fifteen students. The lunchroom and bathrooms were in the basement. The Carnegie school was the pride of the Negro community. It had been named in honor of the Reverend Amos Carnegie, who had come to Marion by 1927 as pastor of Mount Pleasant Methodist Church. When the local school board did not respond to complaints that the school for Negro children was "hardly fit for a stable," Rev. Carnegie took matters into his own hands. He raised money in the Negro community and secured a grant from the Julius Rosenwald Fund, a major supporter of more than five thousand Negro schools across the South. Many of the neighborhood men donated their time and talents to build the school, which opened in 1931. The teachers mostly came from Negro colleges hundreds of miles away, and they brought with them a wealth of talents and culture, such as various languages and musical gifts. They poured their knowledge into children who were thirsty for such information and experiences. A few determined families in the community also found ways to make higher education possible for their children, as my parents had done. Their choice most often was a teacher's training college in Petersburg, Virginia, about 270 miles away.

As a teacher, I made fifty dollars a month, about fifteen dollars less a month than new white teachers with similar training. I had to juggle three different grades, and it was quite a challenge to keep the students all engaged, especially when I was working individually with a particular group. But I worked hard at it. I wanted *all* of my students to feel smart and special. Whether I was teaching them multiplication and division, French words, or the notes on a piano, I used songs, games, and other fun activities to interest the students in learning. I've always loved to learn, and I believe that if students love learning, no one will have to push them to do well. They will want to learn and perform well. A good teacher helps students want to learn. So while my students learned, they also had lots of fun.

Sometimes the school had programs or plays and invited the community to attend. During my second year of teaching at Carnegie, I agreed to direct and play the piano for an all-male comedy as a school fundraiser. The production was a womanless wedding. Some of the parts included men playing women's roles, and they had to wear wigs. This kind of comedy was popular at the time, and for a key role, I needed a man with a good sense of humor and the voice to match. Someone at the school mentioned that Charles and Anna Goble had a musical family and that their son, James, who was home from college for Christmas break, might fit the bill. The Gobles' younger daughter, Patricia, was in my class. Another of the older Goble siblings, Evelyn, taught first through third grades at the school, so I knew of the family. I decided to give this young man, whom everybody in the community called Jimmie, a call. He sounded very pleasant over the telephone when I introduced myself and told him that I was looking for a man to play a particular role in the play.

I asked, "Do you know 'I Love You Truly'?"

He was quite the jokester and said something like, "Why, Miss Coleman, you don't know me well enough to say that to me."

We both laughed. He agreed to play the part in my production, and I was more than a little intrigued by this college man with such a likable personality. When I met him later, at the first rehearsal, he was just as handsome as he was funny and sweet. He was taller than I, but medium height for a man. He also had a medium to small frame, a receding hairline, and the most beautiful brown eyes. We were both pretty smitten right away, and Jimmie and I soon became a couple. His family called him Snook. He was five years older than I and about to graduate from Lincoln University in Jefferson City, Missouri, with a major in science. Jimmie's brother-in-law, Eric Epps (the husband of an older sister, Margaret), was a football coach at the university and had helped Jimmie and two other Goble siblings get jobs on campus and stipends to attend the school. Eric also had served as principal of the Carnegie school before leaving to work at Lincoln. Eventually at least seven of Jimmie's brothers and sisters would graduate from college. The older ones worked and helped their parents send the younger ones to school. That's the kind of loving, helpful family they were.

Charles and Anna Goble had been just teenagers (she was just sixteen) when they married and began having children. The babies kept coming, and the older ones played an active role in helping to raise their younger siblings. They babysat, fed, combed hair, and soothed the cries of the babies while their mother cooked and cleaned. Jimmie was number seven, almost in the middle of the bunch, but to his younger siblings, he was like another parent. While he was away at college, Anna could get the younger ones to straighten up in a hurry with one threat, "Just wait until Snook gets home. He's going to take care of you!" Jimmie adored his family, and the younger ones were so cute and respectful around him. I spent good time getting to know them. The entire family indeed was musical. They entertained themselves by singing in unison many evenings at home when there was little else to do. Several of them could sing and play musical instruments. Jimmie had a lovely bass voice and played trumpet and trombone as well as the tuba and flute. I, of course, played piano and sang soprano. In addition to spending time with his family, Jimmie and I enjoyed taking long walks up and down the Marion hills. We'd walk from church together and hike through the woods. He also enjoyed meeting my family when they visited, and they thought he was a nice young man.

By the end of the school year, I had received a job offer that proposed to take me back to West Virginia. The high school, in Morgantown, offered me $110 per month in salary, more than twice what I was making in Marion. West Virginia already had equalized pay for Negro and white teachers, but that wouldn't happen in Virginia for another year in a battle that would end with a ruling from the US Supreme Court. I was excited about returning home to West Virginia, but I didn't want to leave Jimmie. We were in love, and our relationship was getting pretty serious. One day, he looked into my eyes and asked me to marry him. "Yes," I said, happier than I'd ever been.

My parents liked Jimmie, and I was certain they would give their blessings. But when I called and shared my news, the response was shocking. Daddy did not approve. Crushed and confused, I pleaded for an explanation, but Daddy was not one to explain himself. He could not give his blessing, and that was it. *Maybe he just isn't ready for his younger daughter to get married,* I thought. He'd always wanted me to achieve as much as possible, and maybe he figured I would give up my education to become a wife and mother, as most women did during that time. Whatever the case, he and I had never been on opposite sides before, and his rejection hurt. But he had raised me to think for myself, and my heart was telling me that Jimmie was the one and that this was the time.

In the summer of 1939, I sat down and wrote Daddy a heartfelt letter, telling him how much I respected his opinion but that I was marrying Jimmie anyway. The following November, Jimmie and I had a small ceremony, where we stood before God and our families at the Big House in White Sulphur Springs and pledged to love and cherish one another for the rest of our lives. We celebrated afterward with a reception, where we cut the traditional three-tier, white wedding cake. Despite Daddy's objections, he and Mamà both attended and posed afterward for pictures with us. Many years would pass before Daddy would reveal why he had opposed our marriage.

Although the times were slowly changing for women, Jimmie and I knew that many school systems still did not look too kindly on married female teachers choosing work over family. Unsure whether I would lose my job in Morgantown if we revealed our marriage, the two of us decided to keep our union quiet for a while. Jimmie returned to Marion to teach at the Carnegie School. Then, in the spring of 1940, before we could even live together properly as husband and wife, an unexpected opportunity opened to me.

I was surprised one day after school when I looked up and saw two of my favorite people standing outside my high school classroom in Morgantown. President Davis and Mr. Evans from West Virginia State had come to see me. They had traveled 165 miles or so just to see me? My eyes searched their faces for clues. Was this bad news? Good? They were smiling, so I smiled and calmly invited them inside my room. After a few pleasantries, they explained that they had driven from Institute to talk to me about an important matter. West Virginia University, in Morgantown, was about to become one of the first white universities in the South to begin allowing Negroes to attend its graduate programs, the men explained. The university had been sued by a Negro football player who qualified to attend its law school but had been denied admission. The university could either fight the lawsuit or find a way to comply with the US Supreme Court's decision in a 1938 case (*Missouri ex rel. Gaines v. Canada*) ordering states with schools for whites to provide the same in-state education for Negro students. According to the ruling, states could accomplish this by allowing students of both races to attend the same school or by creating a separate school within the state for Negro students. The case stemmed from a lawsuit that had been filed by Lloyd Gaines, a Negro man who had been denied admission to the University of Missouri's Law School because of his race. The state of Missouri had offered to pay for Gaines to attend law school in a neighboring state. This practice had become common in southern states that did not permit Negroes to attend their graduate schools and did not offer a similar graduate program for Negro students within the state. But Gaines, backed by the NAACP's legal defense team, rejected the State of Missouri's offer and filed a lawsuit, saying the state had violated the "equal protection clause" of the Fourteenth Amendment by providing only white students the opportunity to attend law school inside the state. The case had worked its way up the legal system to the US Supreme Court, which upheld Gaines's claim. The case did not go so far as to overturn the "separate but equal" mandate of the high court's 1896 ruling in the landmark *Plessy v. Ferguson* decision. But legal scholars decades later would come to see the Gaines case as the precursor to the historic *Brown v. Board of Education of Topeka*, which outlawed segregation of public schools.

I didn't know at the time that Dr. Davis, who served on the board of the NAACP Legal Defense Fund, had long awaited the day when the State of West Virginia would have to confront this issue. In 1937 he had talked state officials out of approving legislation that would have granted West Virginia State two million dollars a year over two years to create a graduate program. In a private meeting at the time with West Virginia governor Homer Holt, Dr. Davis told the governor that two separate graduate programs in the state were unnecessary and that the planned four million dollars would be better spent to strengthen the graduate program at the all-white West Virginia University. Dr. Davis said further during his conversation with the governor that the state should make West Virginia State the best undergraduate program it could be so that Negro students then would be prepared to pursue graduate studies at West Virginia University. Governor Holt ultimately agreed. Dr. Davis's strategy, a full year before the Supreme Court ruling in the Gaines case, was about to pay off.

Dr. Davis said that that he had received a call from Dr. Charles Elmer Lawal, president of West Virginia University, who guaranteed that the football player who had filed the lawsuit would be admitted to the law school in the fall without trouble if Dr. Davis could get the young man to drop the suit. Additionally, the state had agreed to desegregate the university's graduate programs without trouble during the upcoming summer session with three of West Virginia State's best graduates. Then Dr. Davis finally got to his point: he had chosen me to be one of the three Negro students to integrate West Virginia University. The other two students were men who were working at the time as principals.

"We've got somebody to pay your tuition," Dr. Davis added.

I was stunned. This was quite an honor and a huge responsibility. I didn't want to let down these two men, who obviously believed in me and were investing so much in my future. Just three Negroes in all of West Virginia were being offered this opportunity. And just one Negro woman: me. I was thrilled, honored, and torn at the same time. This also would mean that Jimmie and I would have to delay further the day when we could live together finally as husband and wife. Jimmie was teaching, and there was no guarantee he would be able to find a job so quickly in Morgantown. But I couldn't say no. I just couldn't. I'd be honored to accept the honor, I told the men.

At least we will not have to move again to another community, I told myself. When I shared the news with my principal, he invited me to stay in his home for the summer. He and his wife would be traveling to Arizona, he said, and their home would be otherwise vacant. I couldn't thank him enough for such a generous offer. My parents, especially Daddy, also were ecstatic that their youngest child would be studying for an advanced degree, perhaps even a doctorate. But they had seen enough in their lives to know that some white residents might not approve. I would be the only Negro woman on the entire campus of West Virginia University. Would I be nervous? How would the other students treat me? Would there be any trouble? This was nearly two decades *before* the Little Rock Nine, when federal troops would have to be brought in to protect those teenagers from frightening mobs as they integrated the high school there. So Jimmie, my parents, and I didn't really know what to expect. Dr. Lawal had assured Dr. Davis that there would not be any trouble, and we put our confidence in that. Still, my parents agreed that Mamà would come from White Sulphur Springs to stay with me and support me during that first summer. On my final day at the high school, my principal, who also worked as an adjunct instructor in West Virginia State's Mathematics Department, had one more thing to offer me, some math reference books that could be helpful, just in case I experienced any problems using the university's library.

Ever in the back of my mind was Dr. Claytor's question of how to become a research mathematician. Perhaps this was the path.

The campus of West Virginia University sat in the Monongahela River Valley on the edges of downtown Morgantown, and it was as beautiful as I had imagined. With its green, hilly landscape and old-style redbrick buildings, the university reminded me a bit of West Virginia State. But the likeness ended there. In the town of Institute, I'd never felt like I stood out. The entire town around our college had been full of Negroes of every hue, from beige to black. That had provided a kind of insulation from the daily indignities associated with race. On the campus of West Virginia University, my insulation was gone. Everybody was white, except the servants, and I rarely even saw them. I met the other two Negro students on the first day during registration. But they went to the law school, as I headed to the Mathematics Department. I don't recall seeing either of them on campus again.

Once I made it to the Mathematics Department, I found my adviser, a white man whose name I've forgotten. He wasn't welcoming, and he didn't seem so happy to see me. After the cursory introductions, he peered at me over the top of his eyeglasses and asked, "What are you going to do with this advanced degree?"

"I'm going to be a research mathematician," I responded.

He seemed amused and asked, "What's that?"

When I explained that I was still trying to figure that out, he asked what job I planned to get in the meantime.

"I'm probably going to do the same thing you do, teach," I said.

He didn't say a word, but his face turned bright red and seemed to say, *How dare you compare yourself to me!*

I hadn't intended to insult him, but it didn't bother me either that he seemed insulted. I hadn't said I was any better than he was, but I knew for sure I was just as good. The white students that I encountered at West Virginia University didn't appear to be bothered by my presence. Many didn't seem to notice me, and some even offered a welcoming smile or nod. There were none of the kind of violent protests that would come in the years ahead with school desegregation efforts throughout the South. I was both relieved and surprised. Jim Crow laws had forced the separation of Negroes and whites with a vengeance for Negroes who dared to challenge the boundaries. My parents had been concerned enough about what might happen that Mamà had come to spend the summer with me. We never talked in any detail about why she had come. But her mere presence and the look of relief on her face each evening when I arrived home spoke silently to me about the potential danger that I faced. I don't know if the other students knew who I was or, given the light complexion of my skin, that I was a Negro. But no one ever called me a nasty name. If anyone had any objection to my presence, they seethed to themselves. I've never been bothered much by what other people thought of me, so I may not have even noticed those in the latter category. What I did notice was that the teachers were fair. They treated me as they did every other student, which is exactly what I wanted. The academic work was never a problem for me, and I met or exceeded the teachers' expectations.

I loved having Mamà in Morgantown with me that summer. We had such a merry time. It felt like the old days, when I returned home from school in Institute, and she was there, often at the stove preparing a hot meal or at the sewing machine, making something lovely. She and I had never before spent this kind of time together, just the two of us, and I grew to appreciate her even more. Even though there was no trouble on campus, it was such a comfort to return home every evening to my mother.

Still, I missed my Jimmie. By the end of summer, I had grown weary of hiding our marriage and spending time together only occasionally. Both of us were ready to live together as husband and wife. Also, Jimmie and I realized that soon we would no longer be able to hide our other beautiful secret: we were going to be parents. Jimmie and I were beyond excited. So was Mamà, in whom I'd confided when I first began missing my monthly cycle.

I knew instinctively that it would be too much to try to juggle becoming a first-time mother with the pressures of being the first Negro graduate student at West Virginia University, so I would not return to the university in the fall. It wasn't a difficult choice for me. The toughest part would be facing the disappointment in the eyes of those who had such high hopes for a different outcome—Daddy, Mr. Evans, Dr. Davis. But all I felt was overwhelming love. Love for Jimmie. Love for our growing baby. Love for the life that we were choosing together.

My heart had told me already what time it was.

chapter 4

THE BLESSING OF HELP

Life changed dramatically for me in late summer 1940. I moved back to Marion and finally was able to live with my husband. We moved in with Jimmie's parents for a while and then rented a couple of rooms in the home of another family. We were excited about becoming first-time parents, and everyone seemed happy for us, even Daddy. Whatever concerns he had about my marrying Jimmie seemed like the distant past. We were all family now, eager to welcome a new little one.

As my small body began to show the pregnancy, family members said I was carrying the baby high, which meant we would have a girl. Sure enough, at 11:40 p.m. on December 27, 1940, I gave birth to a beautiful, seven-pound girl. She was bald, pinkish, and perfect in every way. As planned, I delivered her at Jimmie's parents' house with the assistance of a midwife. Though about half of all women had begun delivering babies in hospitals by then, the use of midwives was still common in rural areas and small towns. Jimmie and I had decided already that if we had a girl, we would name her Joylette, after Mamà. It was the perfect way to honor my mother, who always worked quietly in the background to make sure all of her children's lives flowed smoothly. If one of us needed her, she dropped everything to help, like she'd done when she moved with me to Morgantown to make sure I had the support I needed at home while making history at West Virginia University.

I delved into being a full-time mother and spent my days cooking, cleaning, sewing, and doting on my husband and daughter. Joylette was the embodiment of her name, joy. She rarely cried or wanted to be carried, and she sprang up on her little legs as soon as they were strong enough for her to walk and explore. But while I was enjoying the tranquility of domestic life in the weeks before Christmas in 1941, the Second World War was blasting onto American shores.

Just before 8:00 a.m. on December 7, 1941, the Japanese military stunned the US naval base at Pearl Harbor in Hawaii with a surprise attack. Torpedo bombers and other military planes swooped down from the skies, dropping massive explosives, spraying the battleships docked there with bullets, instantly turning a peaceful morning into a two-hour nightmare. Our huge battleships toppled like toys, sinking into the harbor with thousands of trapped sailors. Before the skies were quiet again, 2,403 Americans were dead, and over 1,000 more were injured. All nine battleships and hundreds of planes and naval vessels there were destroyed or significantly damaged. The largest and once most fearsome battleship, USS *Arizona*, still rests on the shallow floor of Pearl Harbor with the entombed remains of about 900 soldiers.

My family, like most others in the country, heard about the attack the next day on the radio, when President Franklin D. Roosevelt's address to a joint session of Congress was broadcast nationally. He called the day of the attack "a date which will live in infamy." His administration had been in peace talks with Japan, President Roosevelt said. But he offered this assurance: "No matter how long it may take us to overcome this premeditated invasion, the American people in their righteous might will win through to absolute victory."

President Roosevelt asked Congress to declare war against Japan, and within minutes it was done. Germany and Italy, Japan's allies, then declared war on the United States. Our country was officially at war against these so-called Axis nations. We had joined the Allied forces—primarily Great Britain, France, and the Soviet Union—in this new global conflict.

While the Pearl Harbor attack shocked us all, many Americans had been feeling impending doom since September 1939, when German dictator Adolf Hitler invaded Poland as part of his diabolical plan for domination of Europe. He also planned to exterminate the Jews, whom he believed to be inferior. As a matter of fact, Hitler had long been preaching that people of northern European descent were part of a superior Aryan race, and his ultimate goal was to wipe out all others. So this imminent war would be a war against Hitler's brand of white supremacy. After the invasion of Poland, France and Great Britain declared war on Germany, igniting World War II.

It seemed inevitable that the United States would be drawn into war to assist Great Britain in June 1940, when the Germans overtook France. Those suspicions grew stronger the following September, when the United States implemented another mandatory draft, requiring all men ages twenty-one to forty-five to register for possible military service. It was the country's first ever peacetime draft, and young men began leaving their lives and families behind in droves as their numbers were drawn in a national lottery. All of Jimmie's brothers answered the call. My two brothers, Horace and

Charlie, did, too. Horace was deployed overseas. When military officers discovered Charlie's intelligence and writing abilities, he was stationed at Fort Eustis in Newport News, Virginia, as a secretary to one of the military commanders. Jimmie fell into the age group for the draft, but he didn't pass the mandatory physical. He had begun suffering from persistent headaches, but none of us thought the condition was too serious at the time. I was just grateful that my husband didn't have to leave me for the military just months before Joylette was born.

The United States initially remained neutral, a position that most Americans supported, until Pearl Harbor.

After the bombing, I was surprised when my hometown, White Sulphur Springs, became part of a national controversy. More than a thousand diplomats and family members from Germany, Italy, and Japan were living in and around the nation's capital when the United States entered the war. To cut off communications between these emissaries and their home countries, President Roosevelt had them all rounded up and detained at the Greenbrier and other isolated mountain resorts. The Germans and Italians were sent to the Greenbrier, while the Japanese were dispatched to the Homestead, in Hot Springs, Virginia. They had full use of the luxurious facilities, and the Greenbrier's general manager instructed his staff, which included Daddy, to treat these detainees with the same courtesy and respect as any other guests and to provide the same high level of service. That meant these well-heeled detainees would get to swim, shop, play tennis, dine on the finest foods, and have servants at their beck and call. The Roosevelt administration reasoned that by treating the enemy countries' representatives well, American diplomats who were trapped overseas would be treated with the same dignity. But when word got out that diplomats from the Axis nations were being held in such luxury, many Americans were outraged. One railroad executive from New York expressed what many fellow citizens were feeling when he asked in a letter to Under Secretary of State Sumner Welles, "Why coddle German and Jap prisoners who are all bitter enemies of our country, and who would ruin us if they had half a chance?"

White Sulphur Springs mayor William Perry, who also worked as purchasing agent at the Greenbrier, tried to appeal to residents' patriotism when pushed to address their concerns at a town meeting: "We, and I speak for every person in our town, are happy to have this privilege of doing our part during the war crisis. Our whole tradition here in White Sulphur Springs is one of patriotism and support of our government."

But the people could not be persuaded. Newspapers across the county, including the *Charleston Gazette*, opined against the arrangement. When the Japanese diplomats were later added to the Greenbrier detainees the following spring, the town council passed a resolution announcing that it "doth hereby protest most vigorously to the Japanese Enemy Aliens being interned in or near said Town." Negotiations for the exchange of the diplomats lasted until the summer of 1942, when the detainees finally were returned to their home countries. Time ultimately revealed that President Roosevelt's hope for reciprocity on the treatment of American detainees oversees was not realized. News accounts revealed that many of them were treated more like prisoners of war than royal guests.

As the war dragged on, Jimmie and I prayed that God would keep our brothers safe, and we did our best to maintain our lives as normal as possible. Jimmie had bought a car, which was a big deal because only a couple of other colored families in town owned a car. Thanks to my brothers, I knew how to drive it, too, which also was a novelty and the subject of much fascination. Jimmie and I drove to White Sulphur Springs as often as possible to visit Daddy and Mamà. During one of those visits, we met a nice couple who had moved into a rented house across the street from my parents in the summer of 1942. As it turned out, it was Dorothy Vaughan and her husband, Howard, who worked at the Greenbrier as a bellman with Daddy. Dorothy, who was about eight years older than I, had grown up in West Virginia and often walked across the street to swap stories with Mamà. The couple's young daughter also enjoyed sitting on Mamà's lap on the porch, while Mamà read to her. But the family moved away after a short while. A decade later, I would meet Dorothy again, and our lives would become intertwined forever.

For the moment, though, I loved being a full-time mother. On April 27, 1943, Jimmie and I welcomed our second baby girl to the family. She was born at my parents' house in White Sulphur Springs, with a midwife and Mamà taking care of me. We named her Constance in honor of my best friend, Dit. Our little Connie just expanded our capacity to love. At three years old, Joylette took her role as big sister to heart. She would go outside and invite anyone who passed the house to come inside and see *her* baby. Then, to our surprise, before Connie was even one year old, my husband and I learned that our family would be growing again. Jimmie, who grew up in a large family, couldn't have been happier. Daughter number three was born at my parents' home, too, on April 17, 1944, ten days before Connie's first birthday. We decided to name her Katherine, after me, but we called her Kathy. With three growing girls, my life was full. More than ever before, I appreciated the help from both of our families, especially when I had to return to work sooner than expected.

Jimmie had been teaching chemistry at the Carnegie school and was planning to return in the fall of 1944, but he became seriously ill during the preceding summer. He was diagnosed with a condition called undulant fever, an infectious disease believed to have been caused by drinking unpasteurized milk. There had been an outbreak of the disease in Smyth County that summer, with as many as eight cases. For weeks, Jimmie suffered mightily. He couldn't eat much, his fever spiked, and he experienced serious sweats and body pain. When it became obvious that Jimmie would not be well enough to return to school, the principal asked me to fill in for my husband. Jimmie's salary had provided for our family, so I took the job. I became a working mother, juggling the demanding responsibilities of motherhood with those of work—an exhausting balance I would have to figure out for many years to come.

With so much going on in my life, it would have been easy to forget that war was raging on other sides of the world. But our brothers were there, and the absence of so many men from communities across the country left many Americans feeling unsettled. More women than ever had given up domestic life to fill jobs that had been left vacant by their husbands, sons, and brothers who were off at war. Pearl Harbor had snatched the security we'd once felt, knowing that our country had never been attacked by a foreign government on our own soil. We wondered: Could it happen again? Where? How many of our loved ones would have to die before it all ended? And Negro newspapers began raising another question: How could the United States justify fighting Hitler's racism abroad while denying equal citizenship to Negroes at home?

For the longest time, it seemed that the end of the war was nowhere in sight. Despite the devastation at Pearl Harbor, the United States had managed to bounce back. And then came D-Day. On June 6, 1944, about 156,000 British, Canadian, and American soldiers, under the command of General Dwight D. Eisenhower, stormed onto the beaches of Normandy in France in the largest amphibious assault in history. It marked the beginning of the end for our enemies in Europe. By late August, all of northern France had been liberated from the Germans. By the following spring, Germany itself was overtaken. The Allied forces had fought their way through Europe, unraveling Hitler's evil scheme. But the depth of the horror came into focus as the Allied forces discovered mass graves, crematoriums, and piles of dead bodies and ashes of people imprisoned in concentration camps. More than six million Jews were murdered during the Holocaust. At least a half-million other non-Jews also were executed (some accounts say as many as five million), including gay men, people with mental and physical disabilities, Roma gypsies, Jehovah's Witnesses, Catholic priests, Christian pastors, Poles and other Slavic peoples, and members of political opposition groups. The Allied forces liberated thousands of Jewish and other prisoners from the camps, where they had been left to die. As if he could hear the giant footsteps of the Allied forces coming for him, Hitler hunkered down in his Berlin bunker on April 30, 1945, swallowed a cyanide pill, and killed himself with a bullet to the head. Eight days later, Germany officially surrendered.

The fighting between the United States and Japan persisted in the Pacific for a few more months. By then, Harry S. Truman had become the next US president, and his administration ultimately would force the surrender of Japan by using a powerful new weapon in its arsenal: the atomic bomb. On August 6, 1945, the United States dropped the first bomb on the Japanese city of Hiroshima. Five square miles of the city were leveled instantly, and an estimated eighty thousand people were killed. When Japan still did not surrender, the United States meted out more punishment three days later, dropping a second atomic bomb over the Japanese city of Nagasaki. Another forty thousand people were killed on impact. Tens of thousands more people would die later from radiation exposure, including some of our own troops. My family would experience this firsthand in the years to come.

Japan finally announced its surrender on August 15, 1945, ending the deadliest war in history. Since then, that date has been recognized as V-J (Victory over Japan) Day. Even today, the estimated number of people who died worldwide during World War II varies widely, but the National World War II Museum puts that figure at sixty million military personnel and civilians.

The end of the war brought our brothers home, and life seemed a bit safer. Jimmie recovered from the fever, and after that final school year in Marion, he and I moved with the girls back to White Sulphur Springs. Jimmie got a new job teaching high school sciences and coaching football and basketball at the Tazewell County School in Bluefield, Virginia, about eighty miles south of White Sulphur Springs. Because our daughters were so young, he and I decided it would be best for the girls and me to stay close to my parents in a familiar community. That meant Jimmie would have to commute to work, but he was young, vibrant, and willing to do what was best for our family. We rented a house on Church Street, just a few doors down from my parents, and I enjoyed being back in the community where I grew up. Mamà often gave me a break from cooking, and the girls and I joined her and Daddy for breakfast—grits, eggs, fatback (like a thick piece of bacon), and the best biscuits in the world. Sometimes we had dinner and snacks there, too especially when Mamà cooked her delicious applesauce, the girls' favorite. As they grew older, the girls played outside in the yard, and I never had to worry about their safety. Our neighbors on Church Street were like extended family, and somebody's eyes were always watching. At church they'd go from lap to lap, with all of the ladies making such a fuss over my pretty girls. Of course, Mamà and I made all of their clothes.

By 1949 Jimmie was tired of driving so much back and forth, and I worried about him traveling through the mountains on those dangerous, two-lane roads, full of hairpin curves and no shoulders. The sides of the roadways dropped off into deep drainage ditches. Conditions along US Route 60, between Virginia and West Virginia, were especially treacherous during the winter, when drivers had to contend with snow and ice. Once, when the girls and I were riding with Jimmie through the mountains during a snowstorm, the bus in front of us suddenly slid off the highway and dropped into a side ditch. The next thing I knew, our car was sliding off the icy road, too. The girls and I screamed, and seconds later our car landed in the ditch, behind the bus. None of us was injured, thank God. We calmed ourselves, climbed out of the car, and walked over to the bus. The bus driver and passengers all were fine, too. Jimmie and I even left the girls on the bus with a kind passenger as the two of us climbed up the side of the highway and walked a short distance back to a mom-and-pop service station to get some help.

The girls and I moved to Bluefield with Jimmie, and I got a job teaching French, math, and music at the same Tazewell County school where he taught. Each of us made sixty-five dollars a month, which was a decent salary at the time, though not as much as the white teachers were paid. The school was a two-story brick building with the high school classrooms upstairs, where Jimmie and I taught. Our two older daughters' classrooms were on the bottom level, where the first through fifth grades were housed. Joylette was entering the fourth grade; Connie, the first. Kathy, the baby, was not yet old enough to attend school and was supposed to stay home with a sitter. But when she noticed me dressing Connie to go somewhere without her, she whined, "I want to go!" Our "Irish twins" had done practically everything else together for most of their lives, so I brought her along. When we got to school, the principal allowed me to sit Kathy in the first-grade classroom, next to Connie. That became our daily routine. Kathy wasn't officially enrolled in the class, but at the end of the school year, she got a report card and was promoted to second grade. From then on, Kathy and Connie went through school in the same grade.

Our family fit right into Bluefield, a tiny town of about four thousand residents and eight square miles, along the Bluestone River. Some of the world's largest deposits of high-quality black coal could be found there. Just across the state line on the West Virginia side is another Bluefield, with about five times the population. The Norfolk and Western Railway located its headquarters on the West Virginia side in the late 1880s, causing a population explosion. Large numbers of European immigrants and Negroes from the South flocked there via the railroads in search of work in the coalfields. As a result, Bluefield, West Virginia, also had a much larger Negro population—a quarter of the people who lived there at the time. The Bluefield Colored Institute, founded there in 1895, became a popular training ground for Negro teachers, and it was later renamed Bluefield State College.

Though the Negro population on the Virginia side was tiny by comparison, it was a closely knit community. Teachers were respected, and because Jimmie and I were both teachers, everyone loved us and our girls. We never had to worry about finding babysitters. Mothers and teenagers in the community often asked us if they could babysit our girls. There were times, though, when being the child of teachers wasn't much fun. Occasionally the girls had to put up with a bit of teasing. "Gobble, gobble, gobble," one of their peers would blurt, making fun of our last name with turkey noises as the girls walked past a giggling crowd. But we taught our girls to ignore unkind words ("Sticks and stones may break my bones, but words will never hurt me!"). When one of them came home from school in tears because a schoolmate had ridiculed her father's bald head, I wasn't the least bit sympathetic. My response was quite matter-of-fact: "Well, he is bald, isn't he?" I wanted my girls to be tough.

Our daughters also learned the hard way that their father and I didn't play favorites with them in school, and very little escaped our eyes and ears. One day, when it was time for me to go downstairs to teach music to the elementary classes, my eyes immediately fell on the empty seat beside Connie. "Where's Kathy?" I asked the teacher. Well, she said, my baby girl had been caught chewing gum in class and was serving her punishment, facing the wall behind the door. Kathy probably wanted to climb through that wall because I didn't say another word about it during class and left her standing there, missing out on my entire music lesson.

Another time, Jimmie and I gave the girls specific instructions not to follow other kids from the school to a little mom-and-pop store across the street, where the students liked to buy penny candies and cookies during recess and then head down a steep hill to enjoy them. One day, as Jimmie peered out the window of his upstairs classroom, he spotted Connie, midway down the snow-covered hill with her baby sister in tow. He promptly made his way over to the hill, escorted his two girls back to school, and found an empty classroom, where he dispensed some corporal punishment. (That was allowed at the time, by the way.) Some of their schoolmates stood on their tiptoes to peer into the classroom from the hallway through a single clear windowpane on the door. They beckoned Joylette to come and see, too, but when she learned what was going on inside, she kept walking, lest she get spotted and called into the room with her sisters. As far as I know, the girls never tried those little sneaky tactics again.

At work and home, Jimmie and I were a team, and our students were like part of our family. The girls and I rode with Jimmie and his boys on the bus to many football and basketball games, and the boys looked after them in a protective manner, like their baby sisters. I even made the uniforms for the fifteen or so majorettes because the hand-me-downs from the white schools were in such bad condition. For the Homecoming parade through town, our band was quite a sight with about five horn players and a drummer on the back of a pickup truck. There were twice as many majorettes, marching in front. Jimmie and I always worked with what we had. Likewise, he supported my choir performances with whatever I needed him to do. At home both of us pitched in with the cooking and cleaning.

Most summers we supplemented our income by traveling out of town to work for wealthy white families as a live-in maid and chauffeur. One of our jobs was in Rocky Mount, North Carolina, where we worked for the Belcher Family, who had made a fortune in the lumber industry. They lived on a fabulous ranch, and our family stayed in the servants' quarters, a basic apartment on top of their garage. Jimmie and I cooked for them, served their meals, cleaned, and drove them to and from their appointments during the day. These were the kinds of jobs that Negroes took in those days to make ends meet, so I didn't spend any time meditating on the fact that Jimmie and I probably had as much or more education than both of our employers. We did what we needed to do for our family, and kept our peace. The Belchers were kind to us, and sometimes our girls and their girls played together. Once, Mr. Belcher was showing Joylette how to ride a horse, but when he hoisted her on top of the animal and began walking, the horse took off in a full gallop. Joylette, who was about five, slid off right away near a tree and hit the ground hard enough to split open the corner of her mouth. I'm usually pretty calm, but when Mr. Belcher rushed my child inside, her screams and the blood gushing from her lip unnerved me. We grabbed old towels and ice to stop the bleeding, but the cut needed medical attention. The closest hospital was for whites only, and despite Mr. Belcher's efforts, he could not get hospital workers to consent to seeing Joylette that day. Even though Mr. Belcher was able to use his influence to get the hospital to treat her the next day, I was angry. My child was denied medical treatment because of her race. If she had been more seriously injured, what would have happened to her? To this day, Joylette has a faint scar on her lip from that accident. I always appreciated that the Belchers treated us with respect, but at the same time, we understood that we were the help. And we didn't confuse that for a deeper relationship. Nevertheless, when Mr. Belcher's life later ended tragically in a car accident, the news saddened us, especially Jimmie, who had spent more time with him as his primary chauffeur.

About a year or so after moving to Bluefield, Jimmie and I rented a two-room suite in an old mansion that sat so high on a hill that there were probably three dozen steps with landings between to get to the front porch. The place had been owned previously by a rich white family, but a Negro family, the Carsons, had bought the property and lived there with several family members. The house was so large that they took in boarders for most of the second floor. The owners lived on the first floor, with a master bedroom and the living areas—the parlor, living room, dining room, and kitchen. The grandfather of the family had his own bedroom and bathroom on the second floor, where my family's bedroom and bathroom also were located. Our room was spacious, and Jimmie and I slept in a full-size bed on one side of the room, while the girls slept on a three-quarter-size fold-up bed on the other side. The five of us also shared the second floor with another family that lived in a five-room apartment, but we rarely even crossed paths with them.

I have beautiful memories of the girls playing outside on the lawn and rolling down that long hill in the autumn-colored leaves. This was a happy time, but two things happened that would change the world as I knew it.

By early 1950, my brother Horace had developed a mysterious illness. He had been home from overseas a few years since the end of the war, and he began losing weight, feeling extremely tired, and experiencing chills, night sweats, and other symptoms. Horace was always healthy and strong, and it was difficult to imagine him so sick. I was hopeful that Horace would bounce back, the way Jimmie had from the fever, but my brother just kept getting sicker. He was taken to the segregated Veterans Administration Medical Center in Richmond, Virginia, so he could get top medical care. Doctors diagnosed him with leukemia, a blood cell cancer that can be caused by radiation. The news was shocking. No disease was more feared in 1950 than cancer, and with few available treatments at the time, the diagnosis amounted to a death sentence. My family would come to suspect that Horace was exposed to dangerous levels of radiation during the war. Tens of thousands of World War II veterans would develop leukemia and other cancers that appeared to be related to their radiation exposure while serving in the war. They would become known unofficially as the "atomic veterans," and the US Department of Veterans Affairs would develop specific guidelines to determine whether those veterans were eligible for various benefits. Their eligibility was based on where they served and their potential level of exposure. But Horace never got to go through that process. He died on November 7, 1950, at just thirty-eight years old. So much of his life seemed unfinished. My brother had spent some of his prime years as a young man, fighting heroically overseas for his country, yet in some places back home he couldn't ride on the front of the bus or drink from a public water fountain of his choice. Horace had married a wonderful woman named Juhretta, who was from Wyandanch, New York, and at one point worked as a nurse at Freedman's Hospital (now Howard University Hospital) in Washington, DC. But they never had children, and he never got to see the great racial strides this nation would make. He was gone forever. I had never known this kind of pain. I had never lost someone so close, and it all felt surreal. But this was the first of many losses to teach me that even in the worst of times, life goes on. And to keep going with it, you just put one foot in front of the other.

It seems that the ground under my feet had barely settled a couple of years later when one winter evening much of my life went up in flames. Jimmie and I were across town at a wedding reception when we saw people huddled together talking about something. Someone in the group mentioned that the old Carson mansion was on fire. I froze, and my mind began racing: *The Carson mansion? The house where we lived? Our babies were at home!* Jimmie and I looked at one another and dashed out of there. We didn't even stop to get our car. Panicked, we ran the short distance home. As we drew closer, we saw a fire truck parked in front and plumes of thick, black smoke rushing from the windows on our side of the house. Neighbors, including some of Jimmie's football players, were gathered in the yard. Jimmie and I ran frantically toward the house, and I screamed the girls' names. Someone stopped us and told us the girls were safe and waiting for us at a neighbor's house. Jimmie's football players had heard the news, rushed to the scene before we arrived, and rescued the girls from the burning house. The football players had carried our daughters down all those steps to a house a few doors down. Jimmie and I couldn't get to our neighbor's house fast enough. As soon as the girls spotted us, they darted into our arms. I couldn't stop hugging and kissing them.

"Are you okay?" I asked again and again.

Every inch of them smelled like smoke, and they were still coughing and shaking. But I'd never felt so grateful to hold my girls and hear their cries. My babies were okay. Thank you, Lord. Thank you. Nothing else mattered in those first moments.

The fire had been contained to our side of the house, and everyone else also had made it out safely. Another neighbor offered to take my family in for the night, and we appreciated their kindness. We had nothing but the clothes on our backs. We put the girls to bed, but Jimmie and I couldn't close our eyes. Our lives had changed in an instant, and it was truly God's grace and the quick actions of our friends and neighbors that had scooped our girls to safety. When the firemen said it was safe to go inside, Jimmie and I realized that just about everything had been destroyed. What the fire didn't burn, the water and ashes turned into a sooty mess. We had our lives, though. So we picked through the remains and pulled out enough items to fill a couple of boxes and bags. Among the items we salvaged was Kathy's brown baby doll. She had gotten it for Christmas a few years earlier, when toy stores first began selling beautiful dolls that looked like Negro girls. The heat had cracked the doll's face, but Kathy held onto that doll for many years afterward. Despite our losses in the fire, Jimmie and I couldn't feel too sad. It was all just stuff, hard-earned, for sure. Some of the items, especially our pictures, were precious and priceless. But even with our modest teachers' salaries, we could buy new things, take more pictures, and hold in our hearts those long-gone moments that had been captured in black and white. We walked away from the ashes that day knowing we had all we needed to start anew.

I've long since forgotten the date of the fire, what the authorities said caused it, or where it started. But afterward Jimmie, the kids, and I thanked our neighbors and headed home to White Sulphur

Springs. As always, we found open arms and comfort there. Daddy called a white clothing store owner, who opened his business and let us buy some emergency things we needed, and Mamà cranked up her sewing machine. After a few days, Jimmie and I returned to Bluefield, but the girls stayed in White Sulphur Springs, and after recovering from smoke inhalation, they finished the second half of the school year there.

That summer, in August 1952, Jimmie, the girls, and I drove to Marion to attend the wedding of my youngest sister-in-law, Pat. She had been a fourth-grader in my class when I first taught in Marion, but she had grown into a gorgeous young woman. She had just graduated from Virginia State College and was marrying her college sweetheart. Pat was a freshman when she first caught the eye of a handsome senior, Walter Kane, from Big Stone Gap, Virginia. They dated throughout her college years and his graduation and enlistment in the US Army. He had risen to the rank of corporal. Family and friends from both sides had traveled to Marion for the wedding, and we all stayed with the five Goble siblings who had homes there. None of the hotels in town accepted Negro guests, and there were no Negro hotels nearby. Jimmie, the girls, and I arrived a few days early to help out. While busy preparing for the wedding, I asked Pat, "Where's your wedding cake?"

She looked at me sheepishly and said, "Oh, I don't have one."

Being a small town, Marion had no bakeries, and Pat had simply ordered small individual cakes, which was all the local market had to offer. Well, that just wouldn't do. Every bride has to have a three-tier wedding cake. So I talked to Jimmie, and we came up with a plan. We went to the market and bought two regular-size cakes. We cut one down a bit to make the small top layer. Then we found a round hat- box that was the perfect size for the bottom layer. Back then, all ladies wore hats to church, so we had plenty of options. We covered that hatbox in white icing and then stacked the two cakes one atop the other on the box. I swirled the icing around each layer to make it pretty. And voilà! Our baby sister had her three-tiered wedding cake.

The wedding was held at the home of the oldest Goble daughter, Helen, and it was decorated splendidly with lots of fresh white flowers and evergreens. Pat and Walter both looked radiant in their traditional white wedding attire. Everyone was in a celebratory mood, and we ate and danced afterward for hours at the reception. Sometime over the next couple of days, Jimmie and I got word that his grandfather had died, so the out-of-town family members ended up staying in Marion for an extra week. We were sitting around talking one day when another of Jimmie's sisters, Margaret, asked, "Kat, what are you going to do when you get back home?"

"Nothing," I responded, given that it was still summer.

"Well, why don't you come home with me for a couple of weeks," she suggested.

I agreed, and when Margaret's husband, Eric Epps, walked into the room, she excitedly told him about our plans.

"Why don't you bring Snook, too," Eric added. "I'll get you both jobs."

Eric was the former Lincoln University coach who had helped Jimmie and some of his siblings get financial aid to attend the university. He and Margaret had been living in Newport News, Virginia, for years, and they were transporting the newlyweds back to Virginia to honeymoon at the Bay Shore Beach resort in Hampton. The resort, founded in 1898 by several colored businessmen, featured hotels, restaurants, and an amusement park. It was a popular beach getaway for the region's colored families. After their honeymoon, the new couple also planned to make their home in the Newport News area. Among his many roles, Eric served as director of the community center for a huge, federally funded housing development called Newsome Park, where he and Margaret lived. He had been a teacher in the Newport News school system, but he had lost that job because of his active involvement in a class action lawsuit to force the state to pay colored teachers the same as white teachers. Eric was unafraid to speak his mind and was very well connected in the community, so what he said carried much weight.

That evening I told Jimmie, "We're going to Newport News!"

I told him that Eric had said he could get us both jobs. I explained that, according to Eric, a federal agency in Hampton was hiring Negro women with degrees in math. Both Newport News and Hampton were part of the Hampton Roads area, a collection of cities that also included Norfolk, Virginia Beach, and Chesapeake. The area is also known as "the Pennisula." The region was home to one of the country's largest naval bases, five other major military facilities, and was still bustling with job opportunities after the war, my brother-in-law had said.

That night, Jimmie and I lay awake, mulling our plans aloud. We both enjoyed teaching and inspiring young people, and we knew that a loss of two teachers in August, just weeks before the start of the new school year, might put tremendous pressure on the principal to find replacements. But this might just be the fresh start we both needed.

At every point in my life, I'd been blessed by family, friends, and even strangers who stepped in to help me along my journey. Daddy, Mamà, and my family, of course, had been there from the beginning. But there had been so many others: the teachers who recognized my academic gifts at every stage and challenged me, the countess at the Greenbrier who'd connected me with the French chef; Dr. Davis and Mr. Evans, who offered me the chance at graduate school; and the Negro families who had opened their homes and embraced me as a single young professional and then as a young wife and mother, moving from one small town to the next. And then, there was Dr. Claytor, who had opened my mind and ambitions to a possibility I never knew existed.

Was this the path?

The help that my brother- and sister-in-law were offering felt right. Confident that we were making a decision that would change our lives for the better, Jimmie and I looked at each other and agreed: yes, we were doing the right thing.

chapter 5

BE READY

When I was in high school, I learned an important lesson about the power of preparation. In about my junior year, my mother became worried that I had too much idle time. So she walked over to the college campus of the West Virginia Colored Institute and expressed her concern to Mr. Evans, who was head of the Trade and Technical Department at the time. Mr. Evans, who eventually became head of the Mathematics Department, told Mamà, "Send her over to me. I'll give her something to do."

I started working in his office, and he quickly realized I could not type. At the end of that school session, he handed me a ream of paper, a typing textbook, and his typewriter, and gave me these instructions: "When you come back in September, I want you to be able to type."

That summer, I spent much of my time reading that textbook and teaching myself how to type. I knew typing was a great skill to have, and I was determined to be ready for whatever opportunities it opened to me. I practiced daily, getting faster and more accurate, which was extremely important back then because there was no easy, clean way to make corrections. Sure enough, when I returned in the fall, I became the key administrative person in Mr. Evans's office. Then, when I got to college, I landed a coveted job in the president's office and was able to help pay my own tuition. Because I took the time to better myself that summer, I was ready when life offered me a chance to go higher. It was a lesson that would stick with me for the rest of my life.

Nearly three decades later, when the slightest opportunity opened for me to apply for a job that perhaps would allow me to use those high-level math courses I'd studied in college, I couldn't resist. And Jimmie and I took the leap. On a hot, clear day in August 1952, we packed the car and headed with the girls into our future.

As our eight-hour journey wound down to the Hampton Roads corridor, the girls noticed something unusual. Trees were disappearing from the landscape, flashing across their car window frames. When we traveled, they loved looking out the windows at the sky, mountains, and trees. That was their usual backseat view on our road trips from one small Virginia town to the next. But we were entering a new world of wide-open sky, flat highways, buildings where trees used to be, and downtowns with tall, neon signs that brightened the night. For the first time we were moving to the city; and to the girls, it was exciting and a bit scary at the same time.

We moved first to a trailer park on 52nd Street, where a long row of mobile homes sat on the dirt. Ours was a brand-new, compact two-bedroom trailer that had most of the comforts we needed while waiting for a more permanent place. After about a year, my brother-in-law Eric helped us to get off the waiting list and into a three-bedroom, single-story duplex in Newsome Park, the same neighborhood where he and Margaret and Pat and Walter lived. Our home was in the 4200 block of Marshall Avenue, and we literally could look out of our front door at Jimmie's sisters' homes. It was comforting to walk into a new environment with such family support. Despite the newness of our surroundings, the community feeling was familiar. Newsome Park was like my old Church Street community in White Sulphur Springs, multiplied block after block after block. It was a huge, all-Negro neighborhood with families from every income bracket—doctors, shipyard workers, teachers, brickmasons, postal workers, carpenters, entrepreneurs, college-educated, and life-schooled, all mixed in there together. The community of white, look-alike, prefabricated homes had been built during World War II as the federal government's response to a severe housing shortage. Workers from all over the country had poured into the region with their families during the war to fill an abundance of military and defense jobs, especially at the shipyard, but there were not nearly enough decent places for them to live. To meet the demand, the federal Public Housing Administration constructed the largest single defense housing project in the world, as described by news reports at the time, on the East End of the city from 41st Street to 56th Street and from Madison Avenue to Chestnut Avenue. The massive development consisted of Newsome Park, which provided twelve hundred housing units for Negro families, and the similar but separate Copeland Park nearby, offering four thousand homes for white families. Each thrived as small cities unto themselves, as a neighborhood shopping center and an array of businesses built up in and around them. Newsome Park was named in honor of a local Negro attorney and community leader, Thomas J. Newsome, who died in 1942, the year residents began moving into the development.

When we first moved there, Newsome Park was in Warwick, which had just become incorporated as a city to protect its independent status after years of fighting annexation by its landlocked neighbor, Newport News. A city is guaranteed protection against annexation of its territory by adjacent communities. But as Warwick's population growth after the war continued to outpace its tax base and ability to cover the costs of mandatory public services, residents of both Warwick and Newport News years later (in 1958) would agree to a merger. They also decided to keep the better-known name of Newport News, which after the consolidation would become the third most populous city in Virginia. Even before the merger, though, people in Newsome Park always said they lived in Newport News.

We newcomers cared little about the politics of the city's name, though, as we settled into our new community. The girls spent much of their time running around Newsome Park's dirt playground, which had a couple of swings, and monkey bars made of metal pipes. When they saved up enough pennies, they could walk to the soda shop down the street to buy single pickles from a gallon jar. Their new school, Newsome Park Elementary, was a two-minute dash across a few lawns. Joylette was entering the seventh grade, and since she was the oldest child, her father and I allowed her to have her own bedroom. Connie and Kathy, who were going into the fourth grade, shared a room. Their schoolmates and many of the teachers lived in the community, too. The heart of the neighborhood was the Newsome Park Community Center, which had a kitchen and banquet hall, as well as smaller rooms for social and service club meetings. There were basketball and tennis courts, and even a baseball field, where a semipro team, called the Newsome Park Dodgers, played. As director of the center, Eric also managed the team. The games and other events at the center were such fun and an easy way for my family to connect with our new neighbors. Jimmie, the girls, and I made further community connections when we joined our other family members at Carver

Memorial Presbyterian Church. I began singing in the sanctuary choir right away and eventually became director of the youth choir and head of the finance committee. The girls also sang in the youth choir and participated in other youth activities at the church. Jimmie and I didn't believe in our girls having too much idle time, so we filled a good part of their summer days with chores. We did, however, eventually allow them to cross 39th Street to participate in activities at the Dorie Miller Recreation Center. The center had been named in honor of the first Negro to be awarded the Navy Cross, one of the navy's highest honors, for his bravery, manning antiaircraft guns and tending to wounded shipmates during the 1941 attack on Pearl Harbor. Mr. Miller died in battle two years later. The recreation center had the city's only public swimming pool for Negroes. The girls later would take music lessons there in the mornings, tennis lessons in the afternoons, and go swimming. Several members of my new church were members of the Lambda Omega graduate chapter of Alpha Kappa Alpha Sorority, Inc., so I joined the chapter and became even more active in the community. Jimmie also joined the local chapter of his beloved Alpha Phi Alpha Fraternity, Inc.

Eric delivered almost immediately on his promise to help Jimmie find a job at the Newport News shipyard, where he was hired as a painter. Jimmie gladly accepted the position because it was a higher-paying job than teaching, and there was lots of work. Jimmie's education level helped him rise quickly to a position as supervisor, and he was very well liked among the workers. Jimmie liked his coworkers and job, too, and also served as a union leader. Newport News Shipyard was one of the biggest and busiest in the world, given its proximity to so many military facilities. All types of naval vessels, including battleships, were constructed there. Predictions that the postwar season would bring massive layoffs and an economic downturn to the region never became a reality. Shipyard jobs remained plentiful. That was so, in part, because military readiness never lost its urgency as new tensions quickly surfaced between the United States and its former World War II ally, the Soviet Union.

Both countries emerged from the war as superpowers—the United States controlling much of Western Europe and the Soviet Union in control of the East. But their tenuous alliance quickly unraveled, with the two nations jockeying to keep the other's political ideology and influence from spreading. Conflict erupted repeatedly, a time or two even landing the superpowers at the brink of war. Then, in 1949, the Soviet Union tested its first atomic warhead, sending a threatening message that the United States was no longer superior in its military might. Both nations now possessed the powerful bomb technology to obliterate the other. That sobering knowledge may have been the key factor that prevented an all-out physical war. But the heightened tensions and fears during the long Cold War period led both nations to engage in nonstop political maneuvering, threats, and subversive tactics, such as spying, to try to gain an advantage over the other.

Those escalating Cold War tensions also may have created the job opportunities that ultimately enticed me to the Virginia peninsula. In 1950, North Korea, with support and supplies from the Soviet Union, invaded its US-backed neighbor South Korea. The invasion set the stage for a "proxy war," in which the two world superpowers avoided a direct military fight but took opposing sides behind the scenes in this Korean civil war. When Virginia newspapers touted the speed and precision of Russian aircraft after its 1950 attack on a prized US bomber, a shiver must have gone up the hierarchy of the nation's federal aviation program. A headline in the *Norfolk Journal and Guide* read, "Russia Said to Have Fastest

Fighter Plane." The next year the National Advisory Committee for Aeronautics (NACA, the predecessor to NASA) sent Congress a proposal to double its staff agencywide from seven thousand to fourteen thousand in 1953. Military publications trumpeted the job openings, and word traveled through the Negro community with supersonic speed, reaching even the smallest mountain hamlets.

To keep his word about getting me a job, Eric introduced me to the woman who supervised the group of Negro mathematicians at NACA, which was based at Langley Field (now Langley Air Force Base), in Hampton. She and I both were surprised to learn that we had met previously, in West Virginia. She was Dorothy Vaughan—Dot, as she was known by friends and family—the woman who had lived briefly with her husband and children across the street from my parents in White Sulphur Springs. The world suddenly felt smaller, and she and I both marveled at how life had landed us back in the same city after all those years. Dot told me that her agency was not hiring at the moment but that I still should complete an application, and she would look for it.

One September morning a short time later, I drove myself to Langley, completed an application, and began the wait. Between my sorority chapter and church, there were a few ladies who worked at NACA, and they helped me keep track of my application status. Meanwhile I had returned to my other love, teaching. I was working as a substitute math teacher at Collis P. Huntington High School in the East End. Though I was a substitute, I enjoyed helping students understand quadratic equations, geometric construction, logarithms, polynomials, whatever they needed to leave high school with a good understanding of math. It baffled me that so many students disliked math and struggled with it. I figured they had either a parent who didn't like math and told them it was hard or a teacher who didn't have the passion or the patience to make math relevant to their lives. At home I never let my girls tell me math or any other subject was difficult. From the time they were very young, I always tried to incorporate learning, whether it was math, spelling, or creative activities, such as sewing and working puzzles, into their lives. I tried to show them how what they were learning in school connected to our lives outside school. Of course, I had them counting everything—the stars in the sky, the steps from the bottom to the top of the Carson mansion, or the people in church on any given Sunday. On road trips I'd have them add the numbers on the license plates of cars traveling in front of us. Or I'd have them cover their eyes and spell the state. If they were helping in the kitchen, I might write out a recipe, give it to them, and ask them to figure out how much of each ingredient we would need if I wanted to make half of that batch of cookies or biscuits. They got the message early that math is everywhere, and it's not to be feared. Once they learned the basics, they just built on that knowledge with the same confidence. I tried to implement that same philosophy in the classroom as much as was practical. Many years later, I encountered a few of the students I'd taught during that time, and they told me they once hated math but that I'd helped them to understand it as they never had before. That really touched my heart.

In addition to teaching, I also got a job at the Twenty-Fifth Street USO (United Service Organizations) Club as a program director, providing support for military personnel and their families, such as helping them find jobs and housing. The organization's mission is to support these service members and their families in every way possible so they can focus on their primary mission, service to our country. The USO is probably better known for its huge traveling shows, transporting celebrities and entertainers overseas to brighten the days of our deployed troops. But the organization also operated Stateside clubs in large military communities to act as community centers and as hubs for military families and their supportive services.

Eventually I received word that my NACA application had been approved and that my appointment was set to begin in June 1953. Then, twenty-four hours later, the Huntington High School principal offered me a permanent teaching job. I was honored that the principal didn't want to lose me. Of course, the government job offered a much higher salary, and I was surprised when the principal offered to match what I would be making at the other job. I was very grateful, but I didn't have to think about it long. I had moved to the Hampton Roads area to become a research mathematician. And the job at NACA offered me the chance finally to learn just what that was.

The rest of the school year passed quickly, and one day in early June 1953, I was on my way to NACA. My sorority sister Eunice Smith, who also sang in the choir with me at church, offered to drive me to work because Jimmie usually took our car to his shipyard job. Eunice was a delightful woman. She was about five years younger than I, and she had grown up as the youngest of seven children in Portsmouth, Virginia. Her intelligence was evident by the fact that NACA had hired her right after her graduation from Hampton Institute in 1944 with a degree in mathematics. The two of us would become great friends over the years, as our lives intersected in so many ways beyond work. When we made it to Langley Field, I checked in with the human resources director, took care of the required paperwork and other first-day necessities, pinned my employee badge on my blouse, and was told to report to the Aircraft Loads Building. I was about to follow my escort out the door when the director said,

"Don't come in here in two weeks asking for a transfer!"

The director's rudeness surprised me. It was my first day. I was thrilled to be there. Why would I even think about wanting a transfer anytime soon? The remark was condescending and annoying, but I responded politely and kept walking. No one would spoil this moment for me. The classification on my paperwork said I was hired as an SP-3, level 3 subprofessional, basically the entry-level status of most women hired at the agency, regardless of their college degree. But I was a mathematician, a *research* mathematician, and soon I would know what that meant.

I reported first to the office of my supervisor, Mrs. Vaughan, which is what we called Dot at work. Her office was on the ground floor of the Aircraft Loads Building, and I was immediately impressed. She carried herself in a professional manner, and she exuded confidence and authority. I would learn later that, like me, she had started her career as a teacher. She was working at a high school in Farmville, Virginia, when she spotted a flyer advertising job openings in Hampton for female college graduates with math degrees. She applied and to her astonishment was offered a temporary position in 1943. The position offered twice as much pay as her teacher's salary, but her employment was supposed to end in six months. Dot took the job, nevertheless, leaving her children briefly in Farmville with her husband and his family. But her brilliance, meticulous attention to detail, and obvious leadership skills made her higher-ups take notice. Working in her favor, too, was Executive Order 8802, which President Roosevelt had signed in 1941, prohibiting discrimination in federal agencies, including the defense industry. The order also created the Fair Employment Practices Committee, and two years later, an amended version, Executive Order 9346, strengthened the committee and laid out guidelines for it to monitor the progress. Dorothy's hiring status not only became permanent, but also she climbed higher and faster than any Negro woman before her. In 1951 she was named head of the West Area Computing Unit, the all-Negro section of female mathematicians, where I had been assigned. That promotion had made Dorothy the first Negro manager at NACA.

When I first stepped into the room to head to my assigned desk, I was amazed. The office itself was non-descript, arranged somewhat like a classroom with small, professional desks. But I had never seen so many professional Negro women—*mathematicians!*—in one place before. About two dozen of them sat behind their desks, heads down, fingers click-clacking across their desktop calculating machines, making the most beautiful noises. They were called computers, and they were responsible for the tedious mathematical calculations needed by engineers and scientists throughout the agency. There were white computers, too, who did the same work. In fact, the vast majority of computers at the agency were white.

The first female "computer pool" dates back to 1935, when the agency hired four women to perform some of the mathematical equations and hand calculations that beforehand had been done by the engineers themselves. The idea was that this arrangement would free the engineers from spending so much time on essential but time-consuming computing work so they could focus on higher tasks. Despite initial opposition from male staff members at the agency, particularly to spending five hundred dollars per machine for each of the women, the new computers proved that they could do the calculations faster and better.

"The engineers admit themselves that the girl computers do the work more rapidly and accurately than they would," a 1942 personnel memo about the computers said. "This is due in large measure to the feeling among the engineers that their college and industrial experience is being wasted and thwarted by mere repetitive calculations."

The number of computers grew rapidly, boosted by the efforts of Head Computer Virginia Tucker, one of the original four, who also was a former high school teacher with a college degree in mathematics. She traveled to colleges and universities throughout the South, particularly women's colleges, in search of good candidates. The white women's qualifications varied. Some had a college degree, usually in math or science, and some didn't. Many of the college-educated women were former high school teachers. In 1942 there were about 75 computers and 450 engineers among the 1,000 employees at the agency. But World War II dramatically increased the demand for Langley's aeronautics research, particularly aircraft testing, and by 1945 the number of employees at the agency had more than tripled, to 3,220. And Tucker's Computing Department had more than quintupled, to over 400 women in 1946.

By the early 1940s, President Roosevelt's federal antidiscrimination orders were in play, and the agency began hiring Negro computers, who were segregated into the West Area Computing Unit. Many of the early hires came from an obvious source, Hampton Institute, less than ten miles away, but as the word spread, female math graduates from Negro colleges across the country made their way there. Initially the white computers also were organized in a general pool, the East Area Computing Group, but by 1947 their numbers had outgrown their office space, and the group was disbanded. The white computers were then dispatched throughout the agency to work in different sections directly with engineers or research groups. General computing work from the sections or individuals without embedded computers flowed to the West Computers.

Before this day, I didn't have a clue that such a job even existed, but there I was, at the base of some of the nation's most critical mathematical research. The specific tasks varied from one computer to the next, but the majority of the work centered around "reading" film, running calculations, and plotting data on graph paper. For example, when wind tunnel tests were conducted, manometer boards measured pressure changes using tubes filled with liquid. A computer would be assigned the task of "reading" photographic films of the manometer readings, and recording the data on worksheets. The computer might work one-on-one with an engineer or as part of a group assigned to a section to run different kinds of calculations to analyze the data and plot the results on graph paper. The computers did the work by hand with slide rules, curves, magnifying glasses, and the calculating machines to multiply and calculate square roots. The work would then be checked for accuracy and sent to the engineers.

I jumped right in, learning everything I could about the job as quickly as possible. We primarily used two kinds of machines for our calculations. Most of the ladies liked the Friden, but I preferred the Monroe because it was smaller. Dot patiently showed me how to complete the data sheets, based on equations that she or one of the engineers provided. The engineers were in and out of our office all day. One by one they would huddle with Dot at her desk, and after a few minutes exit the room. She would then look up, as if she were studying our faces, and then call one or two of the computers over to her desk to explain their next assignment. Dorothy's mind was the sharpest in that room. She had to understand the math herself first to know the specific skills needed to get an answer. Even more, she had to match personalities, especially when a computer was requested to work on loan for a period of time directly with an engineer in another unit. Scientists can be quite quirky, but Dot seemed to have a knack for figuring out who could work well with whom and get the job done well. She was the mother hen of our unit, the one who challenged, protected, and defended those in her nest, and we all wanted to live up to her expectations.

I'd been there about two weeks when one day Dot looked up

from her desk after one of the engineers left and called fellow computer Erma Walker and me. Erma later would marry a guy named Cartwright Tynes, who worked with Jimmie at the shipyard.

"The Flight Research Division is requesting two new computers," Dot explained.

This was not an unusual request. Computers were often dispatched to a particular unit as the need arose, for a few days, weeks, or even months.

"I'm sending you," she continued.

Every department at NACA had a unique purpose, each critical in some way to the bigger operation. But for an aeronautics agency, Flight Research was the heartbeat. I hadn't expected to be dispatched away from the West Area Computers so quickly, and certainly not to the prestigious group of engineers in Building 1244.

I walked back to my desk and gathered my few belongings. My heart beat fast with excitement, but I wasn't afraid. Dr. Claytor had prepared me well, and I'd done my part to fill in the gaps. The moment I had been anticipating since I was an eager eighteen-year-old girl had arrived. By then I was thirty-four, and as the door to the West Area Computers closed behind me, I was sure of this:

I was ready.

chapter 6

ASK BRAVE QUESTIONS

Mamà was always trying to shush me as a child. I was a meddlesome busybody, and my questions often flustered her. While my older brothers and sister tried studying quietly at the table, I was flitting from one to the other, asking what they were doing and what all those letters on the page meant. Probably just to keep me quiet, Mamà started teaching me how to read and count. It may have surprised her at first how much I could learn at such a young age, but the more I learned, the more she taught. By age four I was reading and doing simple math.

Throughout elementary and high school, my hand stayed in the air, ready to ask the next question. When I was in Dr. Claytor's class in college, the chalk in his hand would be flying across the chalkboard almost as fast as the thoughts flew through his brilliant mind. I usually could keep up, but when I looked around, some of my classmates' expressions appeared as if he were scribbling gibberish. It always baffled me why none of them raised their hands to ask him to slow down and explain. I think they were afraid, as many people often are, of sounding dumb. But I think the smartest people ask lots of questions. I've always loved being around smart people, and the only way to know what they know is to ask questions. When it seemed my friends or classmates in school were too intimidated to ask our teacher a question, my hand shot up, and I asked for them.

One day Dr. Claytor turned the question on me:

"Why is it that you're always asking me questions when I *know* you know the answer?" he asked.

I explained that he had not made the information as clear to *all* of the students as he had to me and that I wanted my classmates to understand, too. That slowed him down, at least for a while. The point is this: if you want to know something or don't understand, ask questions. The path to your destiny may start with a simple question.

So quite naturally, when I landed in the Research Division at NACA in the summer of 1953, I was as inquisitive as ever. I knew I would be surrounded by some of the brightest minds at the agency, and I savored the thought of being able to learn from them. The division chief was an engineer named Henry Pearson, who had started at Langley in 1930 and worked his way up. His office was along a wall inside a larger office with twenty small workstations, filled mostly by white male engineers and a sprinkle of white female computers.

When Erma and I first walked into the office, barely anyone even looked up. That wasn't unusual because people were coming and going through most offices frequently, and the easiest way to avoid distraction was just to keep working. Erma and I found empty desks to await further instructions. I smiled automatically when the engineer in the adjacent workstation turned toward me for what I thought would be a pleasant introduction. But before I could say even a word, he stood hastily and huffed away. Was he upset that a Negro had the nerve to sit next to him? Segregation was still the law in Virginia, and despite multiple presidential executive orders banning discrimination across the defense industry, laws could not legislate hearts. Was he one of the men who opposed women in the workplace? At the time, just about 34 percent of all women worked outside the home, and the percentage was even smaller for white women, who were far more likely than their Negro counterparts to stay at home after marriage. Maybe the engineer didn't want to be bothered with anyone who wasn't at least on his level professionally. Or maybe the guy was just having a bad day. Either way, I didn't react. I didn't try to figure it out. If he had a problem with me for any reason, I would not make it my problem. It was lunchtime, so I opened my brown paper bag and enjoyed my lunch as if nothing had happened. Sometime over the next two weeks, the engineer heard that I was a fellow West Virginia native, and that melted the ice around him. He warmed up to me, and we soon were chatting across our desks. I filed away his initial reaction to me as a distant memory.

I worked well with the engineers. They were full of passion for their work, and they liked that I was interested enough to ask questions: How did you reach this conclusion? Why did you use that equation? What do you expect to learn? Dr. Claytor had pushed me to be more analytical in my approach to mathematics, to understand the whys and ask the right questions to get the right answer. The other computers in the division mostly did their work by the book, hardly ever looking up from their calculating machines to chat with the engineers. Not me, though. I certainly was tedious and thorough in my work, but I also watched the engineers closely. I noticed how they read the newspaper or flipped through the pages of *Aviation Week* magazine each morning to learn the latest industry news and trends. I began starting my day that way, too.

The industry stories I read not only helped me learn more about aviation to connect with my work, but the reading materials also gave me plenty to discuss with the engineers.

By engaging with the guys regularly, I felt comfortable enough to question them when I spotted something that didn't seem quite right. But I knew I had to handle such situations delicately. This was 1953, and many white men were still struggling with seeing women, particularly *Negro* women, in the workplace. I wasn't at all sure how the engineers would react to a woman, a *Negro* woman, a computer, questioning their work. They seemed to enjoy interacting with me when I was in student mode, learning from their brilliance. But would they accept my critiques? Would they accept that the high-level math skills I had attained were at least equal to their own? I had a job to do and couldn't let uncertainty get in the way. So on the rare occasions when I spotted an error, I posed a polite question: "Is it possible you could have made a mistake in your formula?" Of course, I'd already double- and triple-checked the math. One of the things I've always loved about math is that the answer is either right or wrong. And right always prevailed. Accuracy was paramount in our line of work, and the engineers wanted to do good work, so I never had much of a problem with them. They came to respect my questions and rely on my mathematical training and calculations.

Before I knew it, six months had passed, and I was still on loan to the Maneuver Loads Branch of the Flight Research Division. I could hardly believe my luck that my temporary assignment had lasted so long. Only a few of the West Computers had gotten plum permanent positions in other divisions. I had been at the agency less than a year, so I was certain that at any moment I would be sent back to the general West Computing pool. Mr. Pearson might have been content to let me linger in the temporary status a while

Daddy built our childhood home, "The Big House," on Church Street in White Sulphur Springs, which literally was the biggest house on the street. *courtesy of the Johnson family*

My eighth grade classmates and I (white dress in the front) at West Virginia Colored Institute had to dress up for our graduation ceremony. Our teacher, Mrs. Rose Evans, stands behind me in the center. *courtesy of the Johnson family*

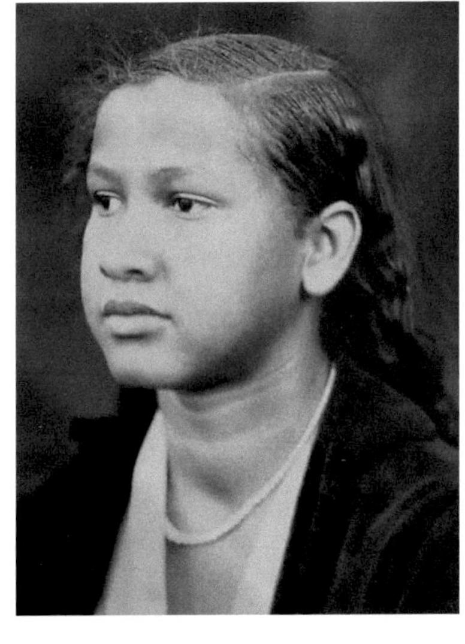

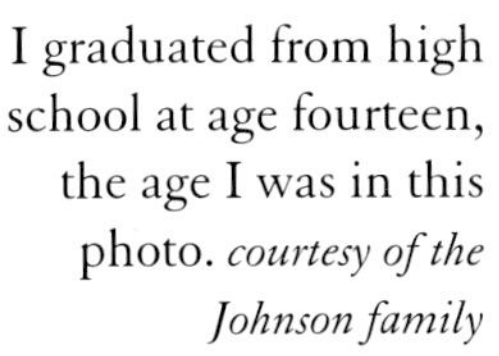

I graduated from high school at age fourteen, the age I was in this photo. *courtesy of the Johnson family*

At West Virginia State College (later West Virginia State University), I (standing far right) was always the youngest girl in the group. *courtesy of the Johnson family*

As a child, I always thought Daddy was the tallest, handsomest man in the world. *courtesy of the Johnson family*

My mother, Joylette Roberta Coleman (Mamà), was a quiet, steady force in my life. *courtesy of the Johnson family*

I graduated from West Virginia State in 1937 with my classmates, Brownie and Jim. *courtesy of the Johnson family*

This is the cast of the womanless wedding play that I produced at Carnegie Elementary School in Marion, Virginia, during my first teaching job. The play brought Jimmie (second from the right) and me together. *courtesy of the Johnson family*

Jimmie and me on our wedding day in November 1939. *courtesy of the Johnson family*

Jimmie and me as newlyweds. *courtesy of the Johnson family*

I taught French, math, and music at Tazewell County High School in Bluefield, Virginia. *courtesy of the Johnson family*

Jimmie snapped this picture of the girls and me on one of our Sunday outings after church at the White Sulphur Springs National Fish Hatchery. *courtesy of the Johnson family*

Jimmie, the girls, and I took this photo as our Christmas card in 1955. It was our last photo together. *courtesy of the Johnson family*

Jimmie and I pose in front of our 1955 Chevy a few months after his brain surgery.
courtesy of the Johnson family

Joylette and me after her high school graduation in 1958.
courtesy of the Johnson family

Daddy stands outside St. James United Methodist Church, which he helped to build in White Sulphur Springs.
courtesy of the Johnson family

This is one of my favorite photos of Jim and me. *courtesy of the Johnson family*

Jim and I took a family portrait with our girls. *Reprinted with the permission of Lifetouch Portrait Studios Inc.*

At eighty-eight years old, I took boxing classes to stay in shape. *courtesy of the Johnson family*

President Barack Obama puts the Presidential Medal of Freedom around my neck, as I sit next to baseball great Willie Mays, who also was honored in 2015. *courtesy of The White House, 2015*

My family and I were honored in 2015 to meet President and Mrs. Obama at the White House (standing behind me, left to right: my grandsons Troy Hylick, Douglas Boykin, Michael Moore, and Gregory Boykin; President and Mrs. Obama; my daughters Joylette Hylick and Katherine Moore; and my granddaughters Laurie Hylick Braxton and Michele Boykin Sanders). *courtesy of The White House, 2015*

I had retired from NASA by the time I got to know astronaut Leland Melvin, but he makes me so proud. *courtesy of the Johnson family*

My daughters, Joylette (far left) and Kathy (center), traveled to South Africa to accept an honorary doctorate degree for me from the University of Johannesburg. *courtesy of the Johnson family*

My family and I, along with NASA astronaut Dr. Yvonne Cagle (in khaki flight suit), prepare to board a private jet to go to the Oscars. *courtesy of the Johnson family*

These are the stunning ladies from the movie *Hidden Figures*: (left to right) Janelle Monáe (who played Mary), Taraji P. Henson (who played me), and Octavia Spencer (who played Dorothy), all on stage with me at the Oscars. *courtesy of the Johnson family*

longer, but for Dorothy Vaughan. She had remained my supervisor, and she knew that with the successful end of my six-month probationary period, I was due a promotion from SP-3 to SP-5 and a salary increase. Ever a fearless advocate for her workers, she paid Mr. Pearson a visit and pushed him to decide one way or the other. I was ecstatic when he offered me a permanent position.

I loved my job in the Maneuver Loads Branch, which was responsible for researching aircraft safety issues. Among my duties as a computer was to review the photographic film from an aircraft's flight recorder, better known as the "black box." I then plotted the data, such as the plane's airspeed, acceleration, and altitude throughout the flight, on large data sheets. I usually did my computing work without knowing any details about the aircraft or the incident in question. But when a small Piper plane fell inexplicably out of the sky while cruising along on a perfectly clear day, I was assigned to help research what went wrong. As usual I was assigned to review the film from the black box. For days I sat in a dark room throughout my entire eight-hour shift and peered down through a binocular-like lens to read the flight coordinates and make notes of the data. The engineers had taught me how to convert the raw data into the metrics they needed, like miles per hour to feet per second. They also provided me the equations to analyze and plot the coordinates on the data sheet to give them a visual image of the flight's fatal path. The engineers then used a test plane to conduct a simulation of the troubled flight, and I assisted in analyzing that data as well. The work was monotonous and strenuous on my eyes, but it was fascinating.

The report, produced from our work, was one of the most interesting things I'd ever read. The engineers discovered that the Piper had flown perpendicular to the path of a huge jet and that the wind stream following the jet was just too powerful for the small plane to withstand. Once a large aircraft passes through an air-space, the troubled air swirls dangerously behind it for as long as a half hour. The small propeller plane had plowed into the turbulent gusts, which sent the aircraft toppling to the ground. Our research helped to change the rules of the airways. New rules were implemented to require minimum distances between planes flying from east to west and north to south to prevent future accidents.

Working so closely with the engineers on these kinds of projects, I developed a good rapport with them. We talked about a range of topics, even school desegregation. Just after the US Supreme Court banned segregated public schools in its historic 1954 ruling in the *Brown v. Board of Education* case, I asked the guys what they thought of the decision. I shared with them what I knew firsthand: that separate schools for Negro children were never equal to schools provided for whites. I talked about the ripped and torn books that were handed down to the Negro students when the white schools got new ones. I had used the well-worn books as both a student and a teacher. I described the dingy, stained majorette uniforms that also had been tossed from the white schools to my black students when I taught high school in Bluefield. The uniform skirts were so shabby that I had to discard them and sew new ones for my students. And I pointed out the pay discrepancies, how Negro teachers got paid less when they often had more education than their white counterparts. But the fundamental unfairness of all those things could not compare to the mental toll of the constant, second-rate treatment on Negro children. To demonstrate this effect during the case, Attorney Thurgood Marshall, then head of the NAACP's Legal Defense Fund, relied on research conducted by husband-and-wife child psychologists Drs. Kenneth and Mamie Clark. The Negro couple had conducted an experiment using four dolls (two white and two painted brown) to ask schoolchildren a series of questions, including which dolls were "nice," which were "bad," which were "most like you," etc. The majority of children preferred the white dolls, and found that the brown dolls were "bad." One child even burst into tears and ran out of the room when asked "which doll is most like you." The tests were used to prove that segregation had engrained in Negro children a feeling of inferiority. When I read about the doll tests, I was even more grateful to Daddy for the self-confidence he had instilled in me from the beginning. After my discussion with the white engineers, the guys all agreed with me that they were in favor of the Supreme Court ruling.

Public school districts throughout the South were slow to respond to the Supreme Court's mandate, and some communities, such as Little Rock, Arkansas, resisted violently. In the coming years, Newport News would comply, and I would be faced with a personal decision about whether to allow my own children to become experiments in desegregation.

In the meantime, Jimmie and I had begun to feel that it was time to move out of our neighborhood. Newsome Park had been perfect for us when we first settled there. We enjoyed living so close to family and meeting new friends, and they all made us feel at home right away. I'd grown so close to Jimmie's youngest sister, Pat, that people often thought the two of us were sisters. When she had her first child, I taught her how to sew so she could make her daughter's clothes. I often took my sewing machine to her home, and we spent hours reading patterns, cutting out fabric, and taking turns at the machine. Every woman should know how to sew. It's one of those old skills that is extremely handy and can save lots of money. I taught my daughters to sew at a very young age.

My girls were at home in our neighborhood, too. But Newsome Park was never meant to be a permanent home for any of its residents. The huge development had been built as a temporary solution to the city's housing crisis during the war, and every now and then city officials would talk about demolishing it. Such talk never ventured past words because the truth was that there still weren't enough homes to meet the demand created by new families constantly flowing into the area. But Margaret and Eric had moved out already and built a beautiful new home in another Newport News community of Negro doctors, lawyers, and other professionals. Their subdivision abutted a whites-only golf course, and on Saturdays the Negro residents knew to stay away from the back windows because golf balls often crashed through their windowpanes. Many of Newsome Park's entrepreneurs also began moving out. A plumber who had lived on our street built a brick home on the edge of Newsome Park, while another neighbor, a barber who was married to a hairdresser, also built a spacious new home in Hampton. After two years Jimmie and I were ready to own a home, too.

On weekends, Jimmie, the girls, and I sometimes rolled through Newport News and Hampton in our car to look at some of the smaller, middle-class subdivisions where Negro families were starting to move. We particularly liked a developing community on a U-shaped street named Mimosa Crescent in Hampton. It was a wooded area that had been mostly swamplands, but we saw potential. Our family physician, Dr. Alonzo W. Douglass, lived at the tip of one end of the U, with his house facing the adjoining street. Mimosa Crescent was mostly empty lots with a few brand-new ranch-style houses toward the front of each end of the street. The houses were affordable, and the opportunity to design our own home just the way we wanted and choose everything from the flooring to the ceiling light fixtures was beyond exciting.

Jimmie and I dug into our savings and purchased a lot on Mimosa Crescent for $1,200—the equivalent of about $11,500 in today's dollars. Our lot was at the base of the U, and two lots were vacant on each side of our property. We began dreaming about our new single-story, three-bedroom house with trees, grass, a backyard, and a patio. One of the girls' school friends even gave us a tiny mimosa tree to plant in the yard later, when we began building. But my family's bliss quickly began to fade when Jimmie got sick with debilitating headaches. Our doctor initially thought Jimmie again had contracted undulant fever, which he had suffered years earlier when we lived in Marion. But the pain and weakness didn't go away. At one point the doctor suspected paint poisoning, since Jimmie enjoyed painting and often helped family and friends when they had a paint job to do. But finally, after months of back-and-forth trips to the doctor and various tests, we were given a diagnosis that would disrupt our world: Jimmie had a brain tumor. And worse, the tumor was in an odd position at the base of his skull. Because of the difficulty reaching it, an operation might be fatal.

The words did not compute. Jimmie and I were a team. We had a dream house to build, three girls to raise, and a lifetime of graduations, weddings, grandchildren, and retirement travel ahead. Frantic questions began flying from my mouth: What did this mean? Was there another way to shrink the tumor? Where could we go to get him the best treatment? Who were the best doctors for this condition? Surely someone out there could help us. There must be another way to treat this thing. There had to be.

We were referred to a white surgeon, Dr. Coppola, who specialized in brain procedures and was said to be the best in the business. This was happening so fast, and Jimmie and I worried that our health insurance might not be enough to cover the exorbitant extra costs of a specialist. But we would do whatever it took. Our family doctor reached out to Dr. Coppola, who asked if he could meet our family. We arranged for him to visit us at home.

Dr. Coppola arrived on a Saturday afternoon in 1955. He was a smallish man, about five feet, nine inches tall with olive skin and dark hair. We were not accustomed to entertaining white people in our home, and wanted to make a good impression on him. I suppose we thought we had to persuade him to care enough to help my husband. So the girls and I dressed up in our Sunday best. I poured him some sweet tea and instructed Joylette, who had become quite accomplished on the piano, to play for him as we sat in the living room. Then I escorted Dr. Coppola to Jimmie, who was lightly asleep in our bedroom. Jimmie's eyes blinked open as soon as I clicked on the bedside lamp, and I introduced him to Dr. Coppola. The surgeon asked Jimmie a battery of questions, examined him, and then walked with me back to the living room to talk. Jimmie's condition indeed was serious, and he needed surgery, Dr. Coppola said.

Okay, I told myself, *there is hope.*

The surgery would be exploratory, and Dr. Coppola said he could make no promises about what he would find or whether he would be able to remove the tumor. This was an intricate procedure that would take about eight hours. *Eight hours*. I just let the words hang there for a moment and braced myself for a fight. I reminded myself: *there is hope.*

Then the financial reality occurred to me again, and I asked how we would we be able to afford this. Dr. Coppola told me not to worry about the costs. He said he was impressed by our beautiful family and would help us. I will be forever grateful to that kind, generous man.

The surgery was scheduled for a Monday in mid-June 1955. Jimmie and I didn't want to worry the girls and decided not to tell them. As soon as school ended for the summer, my sister, Margaret, came to pick them up, and she brought them back to White

Sulphur Springs to spend summer vacation with her and my parents. Jimmie's parents traveled to Newport News to be with us for the surgery.

Jimmie and I slept little the night before his operation but by morning we were calm and resolute. We prayed and drove with his parents just a few minutes away to Riverside Hospital in Newport News. Jimmie's parents and I got to spend a few minutes with him before he was taken away, and we were ushered to a small room to begin the long wait. It was the longest wait of my life. The three of us chatted. We flipped through newspapers and magazines. We prayed quietly. We sat in silence. None of us had an appetite. None of us wanted to leave the room. Finally, eight hours later, Dr. Coppola returned to us. I jumped up, and my eyes asked the question.

"The surgery went well," Dr. Coppola said reassuringly.

Relief washed over me. I asked to see my husband and followed a nurse to the intensive care unit. Jimmie was resting, and he looked so frail with bandages almost completely covering his head, and all kinds of tubes hooked to him. But he was alive, and I was hanging on to Dr. Coppola's words that the surgery had gone well.

When I made it home, I called my parents right away to share the good news. The girls were listening in the background and must have picked up from Mamà's questions that something was wrong. They took turns speaking to me, and each one demanded to know: What's wrong? Where's Daddy? What's wrong with Daddy? There was no keeping it from them now. These were Jimmie's girls, and they loved him fiercely, like I loved my own dad. People always say that there's something special about the bond between a dad and his daughter. He's the first man she loves. It was certainly true for me and my girls.

Mamà told the girls about Jimmie's surgery, and they were inconsolable. They were ages fourteen, twelve, and eleven by then, and they wanted to come home right away. The next morning, my parents put them on a train headed to Newport News. We all were at home when Dr. Coppola called and said Jimmie was awake and that we could see him. The girls, Jimmie's parents, and I hopped in the car and rushed to the hospital. To prepare the girls for what they would see, I told them that their dad was heavily bandaged and very weak but that he would get better. We made our way to the room. I whispered to Jimmie that his girls were there. He blinked a few times and opened his eyes. He seemed to try to focus as he squinted at the faces staring down at him. Then his whole face smiled. Excited to see his reaction to them, the girls instinctively started laughing, clapping, and kissing his face. Their grandparents and I joined in the impromptu celebration. Then a hoarse voice broke through the excited chatter:

"Y'all are too loud, dammit!"

The entire room froze. It was Jimmie, but it didn't sound like him. In all of our years together, I'd rarely heard him utter a curse word. And for him to speak in such an irritated tone when he was so happy to see the girls told us he was in tremendous pain. From then on, sounds of almost any kind bothered him. The clanging from the shipyard across the street. The chatter of his three roommates and their families. The clinking of glass and food service trays in the hallway. No matter how softly we tried to speak when we visited, he winced with our words.

Dr. Coppola was concerned about Jimmie's sensitivity to noise and his susceptibility to infection in a room with multiple roommates. The doctor made a few calls and had Jimmie moved to Mary Immaculate, a Catholic hospital, where he was assigned a private room. According to the hospital's rules, Kathy was too young to visit, but the nurses quickly recognized how important the girls' visits were to Jimmie, and they always allowed his baby girl to slip into the room with her sisters. The nuns prayed with Jimmie, and they also joked with him as his fun-loving spirit slowly returned.

By September 1955, Jimmie had recovered enough to return to Newsome Park. The girls and I were thrilled to have him home. His bandages were off, but the healed incision had left a zipper-like scar from the center of the back of his head to the base of his neck. Fortunately, Jimmie's speech and ability to think had not been impaired, but during his hospital stay he had to learn how to walk again. He walked with a cane, and we kept the house as quiet as a library. The girls rarely invited friends over, but my co-worker Erma, who had become a good friend, became like another aunt to the girls. She occasionally picked them up and treated them to a small shopping trip, ice cream, or a movie. With Jimmie not working, our budget had become extremely tight, and many of the fun activities and extra expenses had been put on hold. Thank heaven for friends.

On his best days, Jimmie mustered enough energy to take a walk outside. He got fully dressed in a jacket, dress pants, and tie and made the girls change from their shorts to something more decent to escort him down the block. He was such a proud man, and it made his day to talk to the neighbors and feel the fall breeze on his face. He even got strong enough to sit in the passenger seat and give instructions to Joylette when we took her out for driving lessons in our black 1949 Chevrolet with a stick shift on the steering wheel. Jimmie sometimes surprised me by standing at the stove to cook dinner, which was one of his favorite pastimes.

In December 1955, the girls' fellow orchestra members at George Washington Carver High School asked if they could dedicate their annual Christmas concert to Jimmie. He and I had been as active at the school as we could be, chaperoning classroom trips, attending football games, and checking with teachers frequently on our girls' progress. I was honored to sit next to Jimmie on the front row in the center, and we beamed as our girls played in the orchestra—Joylette and Connie on the violin, and Kathy on the cello. Those were the most hopeful days, when I could see glimpses of the vibrant young man I'd married and believe that he would be that Jimmie again. But bad days would return. Days when he couldn't will himself out of bed. Days when he seemed to have no more fight. Days when even breathing seemed to hurt.

With the support of family and friends, I managed to take care of Jimmie and the girls without missing a day of work. Then, in late spring 1956, Jimmie returned to White Sulphur Springs for a few weeks with my parents. Mamà, a natural caretaker, insisted on taking care of her son-in-law for a while so that I could rest. I'd tried hard to keep my family's routines as normal as possible and had not realized that I needed the help. Jimmie, once a strong athlete and coach, felt guilty for having to rely completely on me and our family. He even wrote me a letter, apologizing for getting sick and for the stress his illness had caused our family. That's the kind of man Jimmie was. He was the one who was suffering, yet he was apologizing to me.

While he was away, Jimmie also had the chance to visit his family in Marion. Worried that his progress seemed slow, his parents arranged for him to see their doctor. The doctor ran a battery of tests and saw disturbing signs that the tumor was growing back. Jimmie returned home to Newport News, and his doctors verified the diagnosis. There was nothing more they could do. Jimmie was dying. By then it was the fall of 1956, and according to the doctors, Jimmie had just a few more months to live.

How does a wife process that information about the love of her life? I was speechless. The girl with all the questions suddenly had none. But the tears flowed silently. All Jimmie and I could think to do was hold one another until we couldn't anymore. We held hands every moment we could. We locked eyes often, saying everything and nothing at all.

"Promise me you will move the girls out of Newsome Park," Jimmie said to me one day.

I promised.

We couldn't bear to tell the girls right away and waited until Thanksgiving. He was back in the hospital by then, and we wanted them to spend as much time together as possible. At the end of each school day, Joylette drove her sisters to the hospital, and I met them there after my shift at Langley. My pain was one thing, but it was excruciating, watching my daughters leave their father's bedside each evening, never knowing if that would be their last time seeing him alive. When it was time for the girls to leave, they always told Jimmie, "I'll see you later."

They avoided "good-bye," as if it were a bad omen.

On Thursday, December 20, 1956, Joylette, Connie, and Kathy visited their dad at the hospital as usual, and showed him the new eyeglasses they had picked out. He was so weak he could barely respond. My parents had come to town, and Daddy was with me at the hospital, along with two of Jimmie's sisters, while Mamà waited for the girls at home. When it was time for our daughters to leave, they each kissed Jimmie and said, "See you later, Daddy."

By the time the three of them made it home, Jimmie was gone. The moment was as painful and as peaceful as it could be. Mamà broke the news to the girls, and I got home as quickly as I could. Other family members and friends already had started to gather. As soon as I stepped into the house, my eyes searched the room for my girls. They were sitting together on the sofa. All three of them looked shell-shocked as they sat there, sniffling but afraid to cry. Mamà had told them to be strong for me.

"Y'all can cry," I said, opening my arms and pulling the three of them to me. "Come here and cry."

The girls wept in my arms, and I wept inside, summoning every bit of strength I had to hold up my legs and hold back the tears for the privacy of my bedroom.

Friends, flowers, and food started arriving at the house once the word got out about Jimmie. The next day we received a beautiful poinsettia from the girls' school, along with a generous monetary donation that the staff and students had collected for my family. For weeks the support kept flowing in every form—telephone calls, telegrams, cards, flowers, and more.

Two days before Christmas, the girls and I, surrounded by Jimmie's huge family and mine, filed into Carver Memorial Presbyterian Church to bid my husband farewell. Every church pew was packed with heavy hearts—our Newsome Park neighbors, his shipyard coworkers, many of the West Computers, several of our white NACA colleagues, Jimmie's fraternity brothers, my sorority sisters, the girls' schoolmates and teachers, and more. Outside, the heavens wept. Jimmie Goble was well loved.

The days that followed were among the most difficult of my life as my new reality stared back at me. I was a widow at age thirty-eight. My husband was really gone. He would not get better. He would not come home. Ever. At the most unexpected times, I'd see something that reminded me of Jimmie, or I'd have an impulse to tell him something, and reality hit me like an overwhelming wave. The flood of tears would come before I could make it to my bedroom. Jimmie's parents and mine stayed until the end of the year, and they eyed me like hawks. When the wave came one day, Daddy scooped me into his arms. He's not prone to public displays of emotion, and I am my daddy's daughter. But we sat there together, navigating this unfamiliar space.

After a while Daddy broke the silence with a stunning revelation.

"I saw it in his eyes," he said.

I sat up, wondering what he meant.

He'd seen all of this the first time he laid eyes on Jimmie, Daddy explained. Early death for the man who wanted to be his son-in-law. And pain for his baby girl.

It finally made sense to me why Daddy had refused to give his blessing for me to marry the man I loved all those years ago. I had pleaded with him back then to tell me why. Who knows if knowing would have changed anything. I thought about the wonderful life I'd shared with Jimmie and our three beautiful girls. I was grateful that Daddy had done the right thing and kept the heartbreaking answer to my question to himself.

But I knew, too, that I had done the right thing when I followed my heart to Jimmie.

chapter 7

TOMORROW COMES

For the last eight days of December 1956, I hit the pause button on my life. I took time away from work to grieve the loss of my life partner. But I knew that I could not allow myself or my girls to stay locked in the same sad place. Tomorrow kept coming.

When school resumed in January, I escorted Joylette, Connie, and Kathy to Carver High, thanked the staff for their tremendous support, and told the principal that the girls were not to get any pity or special treatment in the days ahead. Jimmie and I had been preparing them for college, and our high expectations would not change. I wanted our daughters to understand that sometimes life hurts, but we have to keep moving forward.

The girls already had chores, but they pitched in even more, helping me with the cooking, cleaning, and ironing at home. They insisted that I must have calculated the exact time it took them to do every chore on their lists to the moment that I walked back into our door each evening. Of course, that is an exaggeration, but I make no apologies. Staying busy is good, and all three of them stayed busy with school and church activities. At school, Joylette was a member of New Homemakers of America, a Negro girls' home economics club that would merge years later with the all-white Future Homemakers of America. She had served as a state officer of the club in the ninth grade and even forged lifelong friendships with some of the other officers. Joylette also was a band member and accompanist for the school choir; Kathy was a cheerleader; and Connie, who had asthma and could not overexert herself, signed up as cheerleader team manager so she could travel with Kathy to the games. All three of the girls attended monthly Presbyterian Youth Fellowship group meetings, led by our pastor's wife, Mrs. Alice Rollins, in the basement of our church. Though the girls sometimes complained that the group was more like an extension of school than a fun fellowship, the young people who participated were exceptional, and they would accomplish great things in the years ahead. Among them were Dr. Carolyn Winstead Meyers, former president of Jackson State University and Norfolk State University, who had been an engineering major at Howard University and was one of the students I mentored; Barbara Starks Favazza, a physician; Cynthia Davis, a jazz singer in Cancún; Eugene Butts, a minister who is now deceased; and a number of educators and other dynamic people.

Since Jimmie and I both were musicians, music was a big part of our girls' lives as well. In addition to playing in the orchestra at school, the three of them also sang in the youth choir at church. The choir director was the musical director at Huntington High School in Newport News, and he taught Connie and Kathy piano in private lessons for a brief while at his home. Joylette had demonstrated an interest and aptitude for the piano as early as five years old, so I began teaching her at home. Jimmie and I also had arranged for her to take lessons from a pianist who performed at the Greenbrier. When we moved to Newport News, we bought a piano our first Christmas there, and by the end of seventh grade at Newsome Park Elementary School, Joylette was accompanying the school choir. The music teacher at Carver High School, Margaret "Peggy" Davis, was so impressed after attending one of the concerts that she walked up to Joylette afterward and said excitedly, "Little girl, I'll see you next year!"

Mrs. Davis saw Joylette's potential and became her "musical parent," nurturing her gifts and pushing her to be better. She was a resourceful teacher, and she exemplified the kind of care and attention that many teachers in the Negro schools gave to schoolchildren to ensure they did not lack cultural and arts exposure and opportunities. Mrs. Davis enlisted Joylette to accompany the Glee Club and the choir and to serve as a student conductor. When Joylette was in the eighth grade, Mrs. Davis signed her up for her first piano competition at Carver High. Joylette made a strong impression on the judges, who awarded her second place, instead of first, simply because they figured she had more years than the others to compete. Afterward one of the white judges, Mrs. Mary Nelson, asked Mrs. Davis about Joylette's training. Upon learning that my talented daughter was not taking private lessons, Mrs. Nelson offered to train her, and she worked with Jimmie and me to arrange a fee that we could manage. Mrs. Nelson lived on Hampton Institute's campus with her husband, who was a German professor there, and Joylette took lessons in their home for the next four years. When Jimmie died, Mrs. Nelson returned to me as a gift every penny that Jimmie and I had paid her for piano lessons. It was such a kind, generous act.

The dedicated Negro teachers didn't let segregation, inferior buildings, and secondhand equipment get in the way of providing our children a top-notch education. They just worked harder to make up for the lack. For example, since Negro children were not allowed to join the Peninsula Youth Orchestra, Mrs. Davis came up with a plan to create her own. In Joylette's eighth-grade year, Mrs. Davis gave an aptitude test to about ten of her top students and formed a string instruments class. She then taught those students over the years how to play the violin, viola, cello, and contrabass. The next two years, she added six to eight more students to play strings and finally added the needed band students who played wind instruments, brass, and percussion. By Joylette's junior year, Mrs. Davis had a sizable group of musicians, which by then also included Connie and Kathy. Jimmie and I couldn't afford to buy the instruments, so Mrs. Davis allowed us to borrow them from her or the school. Just as she had planned, Mrs. Davis had created the first school orchestra—Negro or white—on the peninsula.

To expand the students' musical education, Mrs. Davis required them to attend concerts at Hampton Institute. For one such assignment, students had to attend a five-part concert series, sponsored by the Musical Arts Society. The series was started in 1919 by Robert Nathaniel Dett, a renowned Negro composer, organist, pianist, choral director, and music professor who had been born and raised in Canada. The series included five concerts a year, featuring an array of premier Negro artists, including Duke Ellington, the Alvin Ailey Dancers, and Leontyne Price. The performances were open not just to Hampton Institute students but to the public as well. We paid fifty cents each for our daughters to ride the bus to the Institute for the concerts, and they got to attend performances by operatic tenor George Shirley, who would become the first Negro tenor to perform a leading role at the Metropolitan Opera in New York City, as well as contralto Marian Anderson, who had been the first Negro to perform at the Met, in 1955. Another famed singer who performed during the series was soprano Adele Addison, who had been a former classmate of Mrs. Nelson, my piano teacher, at the Westminster Choir College in Princeton, New Jersey.

One of Joylette's favorite activities with Mrs. Davis was the annual bus trip to a weekend music festival at Virginia State College in Ettrick, Virginia. There the students competed against other Negro school bands, orchestras, and choirs from all over the state and then performed in a concert on the final evening.

When I wasn't shuttling the girls to their events or attending performances to support them, I was going to choir practice or church, or participating in my AKA sorority functions or other community activities. But I also carved out some fun time with Eunice, my dear sorority sister, fellow church member, co-worker, and friend, to attend the Negro colleges' annual Central Intercollegiate Athletic Association (CIAA) basketball tournament. Eunice had been one of the first Negro computers hired at Langley in the late 1940s. She and I were serious basketball fans, and we each used a week of our vacation to travel to the tournament. The event rotated to various cities, Greensboro and Winston-Salem in North Carolina and Norfolk, Richmond, and Hampton in Virginia. The tournament featured sixteen girls' and boys' basketball teams, which competed all week. It was a huge, multicollege reunion, and alumni from all of the participating HBCU colleges packed, wall to wall, into the gym for the games and then spilled out into the town for parties and get-togethers. Eunice and I kept up with the names of all the star players and the team scores, and by the semifinals on Fridays, we had compiled a grid. The crowd grew even larger on Saturdays for the finals. We had a blast each year and always looked forward to attending.

I also made sure the girls and I spent time together. On some weekends we played tennis (or practiced hitting balls across the net) on the courts at a nearby state college. We also made time for sewing, crocheting, and putting together puzzles, and we drove to White Sulphur Springs and Marion to visit my and Jimmie's families as often as we could. The more moments we filled, the less time we had to focus on the gaping hole in our lives. We got through those first months after losing Jimmie by just pushing ahead with the activities that made us feel almost as if our lives were normal.

Then my normal at work changed forever on October 4, 1957, when the Soviet Union launched the first Earth-orbiting artificial satellite. Called *Sputnik 1,* the 22-inch-round aluminum ball weighed 183 pounds and beeped around the Earth every 98 minutes. The launch astonished the world and shattered America's confidence in our technological dominance. By then, we were a decade into the Cold War, and Sputnik felt like a bloodless but humiliating sneak attack that left our scientific community scrambling to strike back. Fear and uncertainty spread throughout the general population as Americans peered into the dark skies night after night in search of blinking evidence that the Soviets were watching us. They wondered: Were we all just targets in waiting as the Soviets marked their path to send bombs or nuclear missiles our way?

Roger Launius, former chief historian for the National Aeronautics and Space Administration (NASA), the agency that would emerge from the panic, described the days after the Sputnik launch this way in a piece he wrote decades later:

> The only appropriate characterization that begins to capture the mood on 5 October involves the use of the word hysteria. A collective mental turmoil and soul-searching followed, as American society thrashed around for the answers. . . .

Almost immediately, two phrases entered the American lexicon to define time, "pre-Sputnik" and "post-Sputnik." The other phrase that soon replaced earlier definitions of time was "Space Age." With the launch of *Sputnik 1*, the Space Age had been born and the world would be different ever after.

Before Sputnik, I had never before spent much time thinking about outer space. Even in the Flight Research Division at NACA, the engineers' projects had been focused on improving flight safety in the earthly realm. But I never again would have the luxury of not thinking about the world beyond the one I knew. And that suited me just fine. We couldn't allow our enemies to learn more about what was out there than we knew ourselves. As I stood with my daughters in our own yard one October night after the launch, I pointed out the satellite, and we tracked it until it slipped slowly out of view. It looked like a moving star, but I was familiar with its path and knew exactly where to look. Standing there, I felt that competitive American spirit rise in me and thought that we couldn't let them get away with this. We've got to do something. Little did I know then that "we" soon would include me. Nor could I have imagined that the maverick Dr. Claytor actually had been prescient when he said I needed one more class to be ready for what lay ahead. He had created that class just for me, and it was the last one that I studied under him: the Analytic Geometry of Space.

While it was shocking that the Soviets had beaten us into space, scientists all over the world, including the United States, had been working collaboratively toward the same goal as part of a worldwide scientific research effort known as the International Geophysical Year (IGY). The IGY was the designated time frame from July 1, 1957, to December 31, 1958, which the international science community had determined would be a peak period to study geophysical phenomena. The White House had announced plans as early as July 1955 to launch an Earth-orbiting satellite for the IGY effort and called for proposals. Two months later, the Naval Research Laboratory's project, Vanguard, was selected to move forward and represent the United States. But by October 4, 1957, Vanguard, whose first satellite was to weigh just 3.5 pounds, was behind schedule, over budget, and had been left in the dust by a significantly bigger Soviet machine.

Before American scientists could even respond, the Soviets launched again, on November 3, 1957, this time sending an even bigger spacecraft, the 1,120-pound *Sputnik 2,* into space with a passenger, a dog named Laika. Playing catch-up a month later, the White House announced plans for a December 6 test launch of a Project Vanguard booster, but the rocket rose just three feet before bursting into flames. A second attempted Vanguard launch, on February 5, 1958, made it four miles before exploding. Both launches were humiliating defeats.

But in the meantime, another unapproved satellite plan that had been developed by a team of US Army researchers for the IGY satellite effort resurfaced and put the United States back in the Space Race. On January 31, 1958, *Explorer 1* was launched successfully from Cape Canaveral, Florida, and made a notable scientific discovery. The spacecraft carried a small instrument, built by physicist James A. Van Allen, to measure radiation encircling the Earth. The instrument verified the existence of the Earth's magnetic field and discovered magnetic radiation belts around the planet. These "Van Allen radiation belts" would thereafter carry his name. A short time later, on March 17, 1958, *Vanguard 1* finally launched successfully into orbit and confirmed the findings of *Explorer 1.*

The success of these launches helped President Eisenhower's administration and the general public breathe a bit easier. But to restore our nation's shattered image as a technological giant, President Eisenhower had to go bigger. He began working with congressional leaders to develop plans for a national agency dedicated to the exploration of space. While there were other proposals, the plan that emerged from the president's Science Advisory Committee was to rename NACA and expand it into an agency whose mission is "to plan, direct, and conduct aeronautical and space activities." Congress approved legislation adopting the plan, and on July 29, 1958, President Eisenhower signed into law the National Aeronautics and Space Act, giving birth to NASA.

Even as the legislation made its way through Congress, the engineers in my division sensed that this was their moment. They and their fellow brainy colleagues throughout NACA were convinced that they knew more about flight than anyone on the planet. They had dedicated their lives and work to studying every aspect of flight and aircraft, and the latest advances in aerospace technology and safety had risen from their research. They were eager to dive in and guide this nation into the Space Age. Clusters of different agency staffers met constantly, trying to learn from one another as much as they could as fast as they could. My supervisor, Mr. Pearson, head of the Flight Research Division, decided to add some structure to the meetings and information sharing. He set up a series of lectures among the two most central divisions, Flight Research and our counterparts in the Pilotless Aircraft Research Division (PARD). Various engineers were assigned individual lecture topics. For example, among the five guys on my team, the assignments included Harold "Al" Hamer, rocket propulsion; Carl Huss, the physics of the solar system; John Mayer, orbital mechanics; Alton Mayo, the issues and challenges objects face returning to Earth; and Ted Skopinski, the math equations to figure out the trajectory (flight path) of a spacecraft.

The seventeen lectures stretched from February to May 1958. I was assigned the task of preparing the data charts and doing the mathematical equations for the lectures. As the engineers prepared for their meetings, they would go over their research with me, and as usual I asked probing questions to be sure I understood what they were doing and what they wanted me to do. The team soon discovered that I could do the equations quicker than all of them. Of the group, I was the last one who had gone to college, and I knew all of this plane geometry. And, of course, Dr. Claytor had taught me that final class, the Analytic Geometry of Space. You see, normally in geometry you learn how things move on a plane. But when you get into the third dimension, you're out there in space, and you have rules. The aircraft has six degrees of freedom when it's out there. So that's another whole science, and it was important to us in that moment to know how to do it. I was working with engineers who would have to figure out, for example, the conditions in space once an aircraft leaves the Earth's atmosphere. Well, I knew how to compute it, and I could tell them faster than they could figure it out because that math was still fresh in my mind. So the task of doing those equations just fell to me. I was thrilled to be able to contribute to this important research.

There was something that bothered me in the beginning, though. It was early 1958, and I knew that the real work of refining, testing, and vetting the engineers' research took place in their group meetings. By then I was working in the Guidance and Control Branch of the Flight Research Division, and I wanted to be in the room with the engineers, absorbing it all, asking questions in real time, not after the fact, as someone gave me the digest version of what happened. So I did what I do when I want to know something. I verbalized the question residing in my mind:

"Why can't I go to your meetings?" I asked the guys.

Silence.

Finally one of them responded, "Girls don't go to the meetings." *Girls*, meaning the computers, meaning me.

"Well, is there a law against it?" I shot back.

Silence.

They knew there was no law. But this had been the practice for as long as there had been female computers. In their minds, their practice was the law, especially since it had never been challenged. But it just made sense for me to be there. I was the one working on the equations. Wouldn't it broaden my perspective to understand firsthand the answers they were trying to get? I didn't argue for myself. I didn't say another word. The silence spoke for me. One of the things I loved about working with the guys is that at the end of the day, everything was about the work. They were smart, logical men. And I knew their minds were silently shooting down every gender- or race-based argument their human nature might have made.

"Let her go," the boss said after some awkward moments.

And just like that, I was in the room.

There were all kinds of meetings, as experts were brought in from various governmental agencies and private industry that had intricate knowledge of flight and space. The engineers recognized that the data they were collecting would be useful down the road, and they decided to compile the data as our own inhouse textbook. Some of the mathematical equations that I had worked out were also included. The book was called *Notes on Space Technology*, and each office was given a copy.

This was such a hopeful time. I was working in my dream job.

And the world—not just the world around me—was changing. Negroes were pushing back against segregation and discriminatory laws that had so long upheld unjust treatment. I'd read with pride a few years earlier about the refusal of seamstress Rosa Parks, a woman just five years older than I, to give up her seat to a white man on a bus in Montgomery, Alabama, on December 1, 1955. I wouldn't learn for decades about the brave, fifteen-year-old Claudette Colvin, who had been arrested for the same in March of that year as she headed home from school. But I was proud of my people there as they galvanized, and for more than a year chose to walk, catch rides, and do whatever they had to do to avoid taking a bus during the famous Montgomery bus boycott. Since Negroes made up about 75 percent of the ridership, the boycott nearly crippled the city's public transportation system, and in December 1956, the month that Jimmie died, the US Supreme Court outlawed segregation on public buses, just as it had done for public schools. But Mrs. Parks and her husband, Raymond, had been fired from their jobs during the boycott and were unable to find work in that city afterward. That's when the president of Hampton Institute at the time, Alonzo G. Moron, stepped up and offered Mrs. Parks a job as hostess at the university's faculty dining room, the Holly Tree Inn. "In this job you have an opportunity to meet many interesting people, for we always have visitors at Hampton," he said to her in a letter. Mrs. Parks accepted the job and arrived on campus a short time later, on September 23, 1957. She worked at the university for a year before returning to Detroit, where she spent the remainder of her life.

At about the time that Mrs. Parks arrived on Hampton's campus, President Eisenhower signed into law the Civil Rights Act of 1957, the first such legislation in eighty-two years. The measure had been designed to protect the voting rights of Negroes, who were being harassed and barred from voting in many southern states. But some civil rights leaders complained that southern Democrats watered down the legislation, rendering it practically useless. Nevertheless, the law created the US Commission on Civil Rights and assigned the US Department of Justice to monitor civil rights abuses. Change was coming. And no matter how hard the segregationists tried they couldn't stop the force of change sweeping the nation.

As NACA prepared to soar into the Space Age, it, too, had to release the baggage of segregation. On May 5, 1958, an administrative memo, announcing the official end of segregation at Langley, circulated among the higher-ups. I hadn't paid attention anyway to one of the most demeaning markers of segregation at Langley, the "colored" restrooms. And I'd mostly avoided the "colored" section of the cafeteria by eating at my desk. But I was relieved for those restrictions to disappear. My emotions were mixed, though, about another of the memo's edicts: the dissolution of the West Area Computers. As a practical matter, only a handful of the Negro computers were left in the general pool and would have to be reassigned. Most of the others, including Erma and me, already were working permanently in other divisions. But I would never forget the emotional impact of walking into that room and discovering this secret sisterhood of smart women whose mere presence as math professionals defied stereotypes. A particular bit of sadness I felt was for Dorothy Vaughan. She was the most brilliant of us all, and I was confident that she would fare well in the transition. But I lamented that her status as NASA's only Negro supervisor would fade away with her unit.

Time kept rolling, and amid all that was happening at work, my oldest daughter was navigating through her senior year of high school. Among the highlights was Joylette's formal introduction to society as a debutante in December 1957. The two-year debutante program was sponsored by a group called the Junior League, which had been formed by a group of mothers from our church and community, including me, to help shape our daughters into refined, respectable young ladies. The debutante tradition was an old, high-society ritual that we wanted our daughters to experience, even though most of us were working and middle-class mothers. The training sessions began in their tenth-grade year, with our daughters learning proper etiquette, everything from which utensils to use at the dinner table to how to dress and comport themselves for formal occasions. They also attended and sponsored community events, such as high teas, to put their acquired skills to use. Then, in their senior year of high school, the club hosted a formal debutante ball, with the girls looking like young brides in their flowing white gowns. Joylette looked radiant in a dress that I made. When Connie and Kathy made their debut three years later, I designed and made their beautiful dresses as well. My brother Charlie stood in Jimmie's place as Joylette's escort, and he did the same for Kathy in December 1960. Connie's escort that year would be the handsome military officer who had come into my life by then.

At the end of May 1958, my family celebrated Joylette's high school graduation. I could hardly believe how quickly the years had passed, and I wished Jimmie could have been there to commemorate those occasions with us. He would have been especially proud as Joylette stood before her classmates on graduation night and gave the salutatory address. I did, however, keep my promise to him. The summer after Joylette's graduation, the girls and I moved into our dream home on the lot that Jimmie and I had chosen on Mimosa Crescent. Daddy, who had come with Mamà from White Sulphur Springs for graduation, stayed to oversee the construction. About thirteen truckloads of dirt had been needed to build up that soggy ground. Daddy took his role so seriously that he even made the construction crew pull up the initial concrete foundation because his precise measurements showed that it was not level. The remaining construction of our new home came together nicely. The girls loved choosing the colors and decorating their new rooms—a salmon shade of pink for Joylette and beige for Connie and Kathy. I chose aqua, which made me think of the ocean. Our part of the neighborhood was very quiet. We lived at the base of the U-shaped street with no immediate neighbors, so when we saw vehicle headlights pointed toward us at night, we knew we were having company.

Then, in no time at all Joylette was headed off to college. Nineteen fifty-eight was proving to be an eventful year, and there was even more.

I was at choir practice one evening, as usual, when a new member strolled into the choir loft and was introduced as Captain James A. Johnson. He had a beautiful bass voice, an easy smile, and a kind spirit. He walked with the straight-back poise of a soldier, but beyond welcoming him to the choir, I didn't show him much attention. The following Sunday, during church service, our pastor introduced the military man to the congregation as a new member and then added, "Ladies, he's single."

Our pastor fancied himself a matchmaker, and I began to suspect that he had done a little behind-the-scenes maneuvering when a short time later at a church picnic Captain Johnson made a beeline to me. His interest in me caught me by surprise. I was a forty-year-old widow with three daughters, and back then under those circumstances, let's just say the chances of finding love again were slim. Plus I wasn't even looking. He was younger, thirty-three, but seemed quite mature and intriguing. He instructed me to call him

Jim; Captain Johnson seemed far too formal. Soon we were chatting nonstop, getting to know one another, and I was blushing like a schoolgirl.

Jim had been born in a small, rural Virginia community called Whaleyville, in the Hampton Roads area. But he had lived in Hampton since he was a boy and considered it home. He enlisted in the US Navy right out of high school in October 1943 during World War II and was shipped to a naval boot camp in Illinois, where he studied propeller repair.

After the war, he attended Hampton Institute on the GI Bill (a benefit that paid for service members to attend college). He joined the US Army's Reserve Officers Training Corps (ROTC) unit there and in January 1951 received his commission as an artillery officer. The following July he was called into active duty to serve during the Korean War, this time as a second lieutenant in the US Army. Thus he had to leave the Institute before finishing his degree. He served on active duty until January 1956, returned to Hampton to finish his degree, and went to work for the US Postal Service as a mail carrier, one of the jobs most frequently available to Negro men returning from active duty. Jim loved being a military man, and he kept his connection as a member of the Army Reserve. For church or formal functions, he proudly dressed in his full military regalia.

We began keeping company, going for walks on the beach, to the movies, dinner parties, and dances. When I was confident that I would be seeing a lot more of Jim, I officially introduced him to the girls. They thought he was very nice.

"Mom, he's cute!" Connie gushed later. That was my Connie, never one to hold back.

I could tell the girls enjoyed seeing me happy and suddenly fussing over my hair, makeup, and clothes again. The next year,

Jim asked me to marry him, but not in the big, romantic way depicted by Hollywood. He was a simple guy. He simply asked. I said "yes." And in a small, private ceremony before God, family, and a few close friends in my home, I became Mrs. Katherine Johnson. We celebrated with a reception in the backyard.

After so much darkness, tomorrow seemed extra bright.

chapter 8

LOVE WHAT YOU DO

I loved my job. I never played sick to stay at home. I never needed inspiration to go to work. My work inspired me. I wanted to be there. I wanted to contribute. I've always believed that to do your best work, you've got to love what you do.

Every day I looked forward to what was next, and I felt fortunate to be close to the action as the nation took its first steps toward outer space. On October 1, 1958, the National Advisory Committee for Aeronautics (NACA) officially faded into history, as the new National Aeronautics and Space Administration (NASA) took its place to chart our country's way into space. Most workers' jobs stayed the same, but there were changes throughout the agency to reflect our added space mission. The name of the Hampton-based facility where I worked was changed to Langley Research Center, and a few other department names were changed as well. New departments were created, new employees hired, and others were shifted into new or expanded jobs, as needed, to advance the mission. The Space Race was on, and the American public was engaged as never before. A poll by pioneer opinion researcher Claude E. Robinson, which was taken just after the Soviets sent both *Sputniks* into orbit, showed that a stunning 95 percent of the public was aware of the launches—and this was long before the twenty-four-hour news cycle and instant access to information. According to the poll, a minority of the public (40 percent) dismissed the *Sputniks* "without serious thought as to what it might mean to them and their country" and the vast majority of them (80 percent) believed America was "at least even" with the Soviets or would "catch up soon." Newspaper editors surveyed for the poll expressed greater concerns.

It seemed clear, though, that the Soviets were dominating the start of the race. By sending a dog into orbit in *Sputnik 2*, they seemed to be advancing toward human spaceflight. With America still playing catch-up, the focus at the new NASA turned to sending the first man into space. This, of course, significantly raised the stakes. Sending a satellite beyond the surface of the Earth was one thing, and even sending a dog into space with no viable plan for its return seemed to much of the public at the time a worthy sacrifice for the sake of scientific discovery. But for this go-round, our scientists knew that it would not be enough just to get a human into space. The agency's best brains had to figure out before our competitors how to get that manned flight back home safely.

Just six days after the birth of NASA, a special panel, led by Langley assistant director Robert R. Gilruth, adopted preliminary plans and schedules for that spaceflight. Mr. Gilruth and his group of about ten engineers had worked practically around the clock to develop the fundamentals of the flight, including the capsule's design, the decision to use existing rockets to power it, and mission control procedures. A Langley veteran who had earned the agency's confidence over decades, Mr. Gilruth would become the quiet force leading the agency into space. He joined Langley in January 1937, shortly after earning a master's degree in aeronautical engineering at the University of Minnesota. He became a leading researcher in transonic and supersonic aircraft and was among the engineers at the old NACA who had shown an early interest in launch technology and human spaceflight. In 1945 he organized an engineering team to investigate experimental rocket-power aircraft, which later became Langley's Pilotless Aircraft Research Division. He headed that division for many years. His research also led to the creation of NACA's Wallops Island launching range in Virginia and successful testing there. Afterward Mr. Gilruth was named assistant director at Langley. Kevin Costner's role in the movie *Hidden Figures* was based primarily on him.

After Mr. Gilruth's panel laid the groundwork for the mission, a new forty-five-member team was created to make it happen. Mr. Gilruth also led that team of thirty-seven engineers and eight secretaries and computers (called math aides after the transition). The assistant project manager was Charles Donlan, technical assistant to the director at Langley. The two men pulled key engineers from all over the agency, fourteen of them from the Pilotless Aircraft Research Division alone, another five from Flight Research, and six others from four other divisions at Langley. Another ten engineers were assigned to the group from Lewis Research Center in Cleveland, Ohio. By the end of November 1958, the team had an official name, the Space Task Group, and its manned spaceflight mission was called Project Mercury. In Roman mythology, Mercury was, among other things, the god of travelers.

Building on the findings of hundreds of previous military studies about the viability of spaceflight, Project Mercury aimed to send a manned spacecraft into orbit around the Earth and to investigate the capabilities of humans to perform in space. The ultimate goal, of course, was to be able to return the space travelers and their space mobiles back to Earth safely. Given that we were already trailing the Soviets, our space team hoped to do all of this by maximizing the use of current technology. There was no time for bold new inventions and years and years of developmental testing.

The Space Task Group decided to call the first space travelers "astronauts," and these men would have to undergo an extensive selection process. First, President Eisenhower specified that the applicants were to come from the pool of military test pilots. That simplified the process somewhat because military pilots had been screened extensively already for national security concerns, and they had significant flight experience. The Space Task Group added additional criteria: The applicants had to be less than 40 years old, under 5 feet, 11 inches tall to fit in the specially designed capsule, in excellent physical condition, and had to have a bachelor's degree or equivalent. In addition to being a test pilot school graduate, they had to have a minimum of 1,500 hours of flying time and be a qualified jet pilot. NASA records show that 508 service records were examined, which narrowed the list to 110 men who met the criteria.

The number was reduced further to 32 men, who went through individual interviews, physical and psychological examinations, and training exercises. Finally, the remaining 7 astronauts were introduced at a press conference in April 1959, and they became instant celebrities. They were the walking embodiment of our nation's space dreams. They were willing to risk their lives to travel to the unknown and push the boundaries of human exploration.

"These men, the nation's Project Mercury astronauts, are here after a long and perhaps unprecedented series of evaluations, which told our medical consultants and scientists their superb adaptability to their coming flight," NASA administrator Dr. T. Keith Glennan said at that press conference. "Which of these men will be first to orbit the Earth, I cannot tell you. He won't know himself until the day of the flight."

It was so exciting when those of us behind the scenes occasionally ran into the astronauts at the office, but they were just as excited to meet us. They were assigned to the Space Task Group and each had a job to do: Scott Carpenter was assigned to handle communication and navigation; Gordon Cooper served as liaison with the team developing the launch systems; John Glenn worked on cockpit and instrument panel design; Gus Grissom developed the control systems; Wally Schirra worked on space suits and other life support; Alan Shepard worked on tracking and capsule recovery; and Deke Slayton worked to integrate the Mercury capsule with the rocket that would boost it into space.

The rest of the Space Task Group was organized into divisions, with each responsible for a different element of the flight. John Mayer was the first of my officemates to be transferred there. He was assigned to the Flight Systems Division, headed by Maxime A. Faget, the engineer who had designed the blunt-ended, cork-shaped capsule chosen to carry the first man into space. The group worked in a separate building, but John regularly made his way back over to Flight Research (renamed the Aerospace Mechanics Division) to get computing help from Carl Huss and Ted Skopinski. As usual, Ted brought me in to assist. I had worked closely with Ted as his computer, and he trusted my mathematical and analytical skills. We spent many hours working through the equations, analyzing the math, drawing diagrams, and going back to the drawing board again and again.

It had been decided from the start that a rocket booster would fire the ballistic capsule into space, like a bullet shot from a gun. For the initial suborbital flight, the capsule would go up into space, turn, and head right back down for a landing in the Atlantic Ocean. US Navy ships would be on standby to pull the capsule and our first space explorer out of the water and back to safety. Project Mercury's ultimate goal, though, was to orbit the Earth, and that trajectory—a circular path around the planet—was far more complex. Ted knew more than practically anyone at Langley about how to figure out the trajectory for this. So he was tasked to write the report that would lay out the path of NASA's first orbital flight. I wanted in on it, too.

"Let me do it," I suggested to Ted. "Tell me where you want the man to land, and I'll tell you where to send him up."

He didn't hesitate to let me run with it. We collaborated on the report, but much of the math was my work. So when Ted and Carl left our division to join John in the Space Task Group, Ted suggested to our boss, Mr. Pearson, that I should complete the report. Mr. Pearson wasn't the most progressive fellow when it came to women in the workplace, but he agreed to let Ted move on and for me to finish what we had started. Just after Thanksgiving in 1959, I handed over the thirty-four-page report, full of all kinds of equations, a couple of launch case studies, tables with sample calculations, lots of charts, and reference texts. Entitled "Determination of Azimuth Angle at Burnout for Placing a Satellite over a Selected Earth Position," the research report went through the agency's usual detailed process of review, analysis, and revisions before publication in September 1960. I was extremely proud of the work Ted and I had put into the research and even prouder that our work would be used to direct NASA's first flights into space. So I had signed my new name—Katherine G. Johnson—to the report, just like the men. It was a rather assertive move at the time because women just didn't do that, even though they often contributed significantly to the work. When the report was published, it was the first time a woman of any race in my division had been listed on the research paper as a coauthor.

By the time the report was published, Jim and I were adjusting to life as newlyweds. He folded easily into my family, and the girls really liked him. He didn't instantly try to take over and assume the role as their father. That wasn't his way. He recognized that the girls and I had experienced the most devastating loss of our lives, and that while I had opened my heart to love again, it might take time for the girls to open their hearts completely to a new father figure. Jim was a good, patient man. The girls and I particularly enjoyed shopping for him. As a military guy, he hardly possessed any dress-up clothes aside from his formal military uniform. In a house full of women who enjoyed getting dressed up to go out, we wanted to vary his wardrobe a bit. It gave us an excuse to shop.

In the fall of 1960, Joylette was in her junior year at Hampton, and Kathy and Connie were entering their senior year of high school at Carver. We had decided as a family that they would remain there to protect them from the chaos in schools across the state over court-mandated desegregation. Virginia governor Lindsay Almond shut down schools that were trying to comply with the US Supreme Court's order to desegregate. Just across the water in Norfolk, about ten thousand students, more than half of whom were from military families stationed there, had nowhere to go. Schools in Front Royal and Charlottesville were closed, too. The schools in Newport News remained open because they remained segregated. Moving to the house on Mimosa Crescent had changed the school zones, putting us in the all-white Hampton High School district. But the school system was so determined to maintain the status quo that Negro families in white school zones were essentially paid three hundred dollars per child to send their children to Negro schools. The payments were disguised as "school fees," but a similar practice of granting "scholarships" to Negro students to attend integrated white colleges and universities outside of their segregated states also enabled my sister Margaret to get a master's degree from Columbia University and brother Charlie to earn his master's from New York University. The states where they were living at the time—Sister in West Virginia and Charlie in North Carolina—paid their tuition at those prestigious New York universities.

Connie and Kathy had wanted to stay at Carver anyway, and I was comfortable that they were getting a great education there from well-trained teachers, who always had their best interests at heart. Unlike when I had become the first Negro graduate student at West Virginia University decades earlier, the state of Virginia was actively resisting integration and whipping up a frenzy among residents in the process. Once I'd seen what those Negro teenagers experienced in Little Rock, I couldn't unsee it. And I was content that my daughters were safe right where they were.

Connie and Kathy graduated from Carver in June 1961 and for the first time went their separate ways. Kathy chose Bennett College for Women, a United Methodist women's college for Negroes in Greensboro, North Carolina. Connie headed just a few miles away to Hampton Institute, where Joylette was entering her senior year. By then, Jim was already volunteering as alumni adviser to the army's ROTC unit at Hampton Institute, a role he had assumed in February 1956, the month after he returned from active duty. He loved guiding young people who were planning a military career, and he jogged around campus with them once a week to stay in good physical shape. Jim spent almost as many hours on the Hampton Institute's campus as paid faculty and staff. His presence there was a relief for me because when Joylette had first left for college a few years earlier, he was able to keep a distant eye on her. She had been first to leave the nest, and like any mother, I worried about her. Just as she had started to settle into college, Negro college students across the South were inserting themselves into the growing movement for civil rights and racial equality. I know my girls and just sensed that Joylette would want to be part of it all.

Joylette was a sophomore on February 1, 1960, when four Negro students from North Carolina A&T College in Greensboro sat down at a "whites only" lunch counter at the downtown Woolworth store. The four guys were all college freshmen who had been inspired by Dr. Martin Luther King Jr. and came up with their own plan to do something to combat segregation at the store. When a white waitress refused to serve them and asked them to leave, they politely refused and stayed seated for about an hour until the store closed. To their astonishment, they were not arrested. The next day, the four recruited more participants, and about two dozen students returned, were again refused service, and sat at the counter for about four hours. This time the protest drew major media attention. By the third day, the students had formed the Student Executive Committee for Justice and sparked a movement.

The sit-ins began flowing like a wave to other cities in North Carolina and beyond. Students at Hampton and other Negro colleges began plotting, and Joylette mentioned to me that she wanted to participate. I discouraged her from joining the movement. That may not be popular to admit today, but at the time I was looking at that situation as a mother, concerned about my child's safety and future. If Joylette got arrested, that would make finding employment after college, especially at NASA, very difficult. She had decided to major in math, like me, much to the disappointment of her high school music teacher, Mrs. Davis. Mrs. Davis had hoped that the many years of piano and organ lessons, competitions, and her musical gift would inspire Joylette to choose music as her major. I had not influenced her decision, but once she chose math, I couldn't help thinking of how exciting it would be for her to join me someday at NASA. All of that hope might vanish if she got could not pass the background check, and I expressed this to her. Nevertheless, just ten days after the first Greensboro sit-in, students from Hampton Institute became the first Negro college students outside North Carolina to conduct a sit-in at the local Woolworth. They endured the rudeness of the white waitress and the taunts of white customers and kept going back. Like so many of her peers whose parents had some misgivings about the sit-ins, Joylette joined a group of her friends and participated in one of them anyway, without telling me. Years later, she would open up about the terror she felt inside as she sat quietly sat at the counter, trying to pretend she didn't hear the crowd of angry white hecklers calling the students derogatory names and screaming at them. But she felt gratified for being part of a movement that ultimately would desegregate lunch counters, restaurants, and department stores across the American South.

Two years later, Kathy also joined the protests in Greensboro in her sophomore year at Bennett. She had missed the first wave of protests in Greensboro because she was still in high school, but those 1960 sit-ins had ended the summer after they began with the integration of lunch establishments inside a few downtown department stores. But by the time Kathy arrived on Bennett's campus, students were growing frustrated again with the lack of more widespread progress. In 1962, students from North Carolina A&T and Bennett in Greensboro started a new round of protests, targeting about a dozen businesses that remained segregated. This time the protests were coordinated by a local chapter of the Congress of Racial Equality (CORE), led by two North Carolina A&T students. Once again I tried to talk my daughter out of participating. And once again my daughter followed her own heart. She attended meetings where students were trained in civil disobedience, how to walk in picket lines with a guy between two girls, and how to continue marching and looking straight ahead, no matter what the mob of bullies that may surround them said or did. The freshmen and sophomores initially were allowed to play only supporting roles, for instance making sure the upperclassmen who were participating got their homework turned in, even if it meant typing papers for them. But Kathy said that when she finally had a chance to walk on the picket lines in front of the five-and-dime store in Greensboro, she "didn't have sense enough to be scared." That changed one day, though, when she and her roommate, Pat, somehow got separated among the protesters, and Pat ended up going to jail with the group of students who were arrested. Kathy later confided to me that she felt terrible about it because the girls had promised Jim and me, as well as Pat's parents, that they would always look out for one another.

The president of Bennett, Dr. Willa Beatrice Player, was a fascinating woman who played a critical role during the protests. She urged her "Bennett Belles" to follow their convictions if they wanted to take a stand for justice. She had become the first Negro woman named president of a four-year institution when she was inaugurated to the post in the fall of 1956. Nearly two years later, when the NAACP and a few Negro ministers invited Dr. King to speak in Greensboro just after the Montgomery bus boycott, some Negro churches and universities were reluctant to attract controversy by hosting him. But Dr. Player stepped up. "Bennett College is a liberal arts college where freedom rings, so King can speak here," she announced. He later spoke to an overflow crowd in the Annie Merner Pfeiffer Chapel there on February 11, 1958.

Kathy remembers Dr. Player as a pretty, prim, and proper lady who always wore dark suits or dresses—the custom for all of the

Bennett Belles at the time—and her hair pulled back into a bun. She had such a commanding presence that Kathy declares the yard parted like the Red Sea when she walked across campus from the president's residence to the academic buildings. Greensboro police began arresting students in mass numbers between November 1962 and May 1963, and Dr. Player boldly stood up to them. On May 17 and 18 that year, about seven hundred protesters, mostly students, were arrested while picketing at several businesses that had rejected an integration proposal by the Chamber of Commerce and Merchants Association. The prisons filled quickly, and many of the protesters were bused to and held in an old polio rehabilitation center that had been condemned. Dr. Player demanded to see the students and defended their actions to faculty, staff, and students during a Friday chapel service. She explained that she had seen her girls and called their parents. She visited her jailed students daily, and while the city struggled to feed the masses of citizens arrested during the protests, Dr. Player helped arrange the distribution of mail, food, and class assignments to her Belles.

About two days after Chapel, when the students found out that their schoolmates were being held in a moldy old building, they arranged for a bus to take them there. The bus got as close as it could, and the students walked the rest of the way. They then surrounded the old building, and a young Jesse Jackson, then president of the student body of North Carolina A&T, led them in a prayer vigil. State troopers stood guard, facing the students. Before praying, Jesse appealed to the officers, "We serve one God, will you please remove your hats." Some of them reluctantly complied.

The students who were jailed inside appeared at the windows and began shouting their names with the hope that their classmates would send word to their families that they were okay. Kathy was the one who called Pat's parents, who lived in Houston, to tell them about her arrest. I'm sure that was one extremely difficult phone call. The federal government eventually stepped in, and the students were released.

Of course, I knew none of these details at the time. Television news coverage then wasn't nearly as widespread as it is today, so there was no way for me to know, until Kathy told me later. I suppose she didn't want me to worry. But Kathy remembers fondly Dr. Player's bravery and dedication to her students.

In the middle of the student protests, NASA inched closer to human spaceflight with flight tests for Project Mercury. But one delay after another pushed the projected launch year from 1960 to 1961. Part of the challenge was that the engineers had to build a communications network that would enable the spacecraft to stay in contact with Mission Control in Florida as it orbited the globe. That meant building eighteen communications stations in different parts of the world so that the astronaut would never lose touch.

On January 31, 1961, our space team launched into space a chimpanzee, nicknamed HAM for Holloman Aerospace Medical Center, the New Mexico–based facility that prepared him for the flight. The suborbital flight (up into space and back down) was boosted by a Mercury Redstone rocket launched from Cape Canaveral, Florida. Of all the previous test flights, this one was particularly crucial because the chimp's physical structure was most similar to a human's, and the animal had been taught to perform certain tasks in the spacecraft to test whether the weightlessness of space negatively impacted its ability to function. The flight also gave engineers a chance to work out any possible technical problems. The flight lasted sixteen minutes and thirty-nine seconds, and there were a few technical issues. The spacecraft moved faster and shot higher than expected. The capsule splashed down in the Atlantic Ocean far off course and was filling with water by the time a rescue ship pulled the passenger to safety. HAM was still healthy and had been able to perform the basic functions he had been trained to do, which signaled that space had not disabled him. NASA sent up one more test flight without a passenger on March 24, 1961. Finally, we as an agency were ready. Before we could make our big move, though, the Soviet Union shot ahead of us again.

On April 12, 1961, Yuri Gagarin, a Soviet Air Force pilot and cosmonaut, became the first person to enter space, on a 108-minute flight that also circled the Earth. I knew it must have been crushing to our astronauts, who had been working so hard for the moment when one of them could turn the momentum in our country's favor. My concern from the beginning of NASA was that the agency shared everything with everyone, and so the world knew what we were doing. However, as an agency we didn't know what everyone else was doing, and once again we were blindsided. Fortunately, this time we weren't far behind. Astronaut Alan Shepard, a US Navy test pilot and World War II veteran, was selected to become the first American in space, and on May 5, 1961, just three weeks after Gagarin's flight, he took that journey before a live television audience. Millions of television viewers tuned in as NASA's first-ever live launch broadcast showed the Mercury capsule boosted into space by a Redstone rocket for a suborbital flight that lasted fifteen minutes and twenty-two seconds. Shepard had named his tiny capsule (just six feet, ten inches high and six feet in diameter) *Freedom 7* to honor the nation's first seven astronauts. NASA called the mission Mercury-Redstone 3 (MR-3). The capsule splashed down in the Atlantic Ocean about three hundred miles from its Cape Canaveral launch pad, and a rescue helicopter pulled it and Shepard from the water.

Decades later, in a 2007 biography of his life, Shephard expressed his disappointment that the Soviets made it to space first.

"We had 'em by the short hairs, and we gave it away," he is quoted as saying.

Though much shorter and less complicated than the Soviets' first orbital flight, our first human spaceflight at least put us back in the race. Afterward I felt a huge sense of relief. I knew my numbers were right, but any small, technical thing could have gone wrong and caused a disaster. I had done my job well—calculating the trajectory and what is called the "launch window," the time frame during which the launch had to occur to be successful. In case of electronic failure, I'd also plotted a backup navigation chart for the flight. We worked as a team behind the scenes, and no one's job was more important than the other. Little did we know in those first moments after the success of MR-3 that the space program was about to get a much-needed jolt.

The water was barely dry on the Mercury capsule when new US president John F. Kennedy made his historic speech about the space program on May 25, 1961, to a joint session of Congress.

"Now it is time to take longer strides—time for a great new American enterprise—time for this nation to take a clearly leading role in space achievement, which in many ways may hold the key to our future on Earth," President Kennedy said.

He applauded Shepard and then asked Congress for the funds to support an ambitious new space goal, which he laid out next:

"First, I believe this nation should commit itself to achieving the goal, before this decade is out, of landing a man on the moon and returning him safely to the Earth. No single space project in this period will be more impressive to mankind or more important for the long-range exploration of space; and none will be so difficult or expensive to accomplish."

The moon? President Kennedy was right about the difficulty part. Our space team had not yet even caught up to the Soviets with a successful orbital flight, yet he was already pushing us to the moon. The president asked Congress to approve a whopping seven billion to nine billion dollars in funds over the next five years to accomplish that goal. It soon became clear that President Kennedy's monumental vision was bigger than Langley could handle. Congress approved the funds to build the new Manned Spacecraft Center, and in August 1961 the official search for the Space Task Group's new home began. Though Langley officials campaigned to keep the heart of the space operation, nearly two dozen other sites were considered. Houston ultimately got the nod. Historians later would point to the influence of some powerful Texans, including the one who sat at the president's right hand, Vice President Lyndon B. Johnson, and Speaker of the House Sam Rayburn. When President Kennedy was assassinated on November 22, 1963, Lyndon Johnson ascended to the presidency, and a decade later, the Houston space center would be named in his honor.

My former officemates Ted, John, and Carl all followed the Space Task Group to Houston, and I was asked to join them. Working with those guys had been exhilarating, and I gave the move some serious thought. But when I talked to Jim about it, we both concluded that we could not move so far away from the girls and our extended families, who all were no more than a few minutes or a few hours away by car. The Newport News area had become my home, and for the foreseeable future, it would stay that way.

Even as the Space Task Group made plans to move, the group members never stopped working. And neither did the rest of us. My days of working a straight, eight-hour shift and making it home in time for dinner every night went away with NACA. I was spending longer hours at work, and sometimes I'd even have to return to work after dinner. I appreciated that my husband, as a military man, understood not just the secretive nature of my work, but also the long hours of commitment. Just two months after the first suborbital flight, our space team launched a second one with astronaut Virgil "Gus" Grissom, a US Air Force pilot and war veteran, piloting the spacecraft. The flight, on July 21, 1961, was officially dubbed *Mercury-Redstone 4*, and Grissom nicknamed it *Liberty Bell 7*. After fifteen minutes and thirty seconds, the flight landed in the Atlantic Ocean and almost ended in disaster when the capsule's hatch blew off. We all breathed easier when a helicopter pulled Grissom out of the water. The capsule, which already had begun filling with water by the time the rescue effort began, was too heavy for the helicopter to lift and could not be salvaged.

With every flight, every mistake, every correction, every test and retest, Project Mercury moved closer to its ultimate goal of manned orbital flight. Meanwhile, another transition was in the making, and it would significantly change my world: the growing use of electronic computers. When NACA purchased its first "electronic calculator" from Bell Telephone Laboratories back in 1947, that should have signaled what was to come. The huge machine took up an entire room and noisily coughed up answers to the engineers' equations for transonic flight research. But its advantage was as commanding as its size: speed. In just a few hours, the calculator could perform a task that took humans a month to complete. Plus it could work all night without a break.

A few years later, in the early 1950s, NACA purchased its first IBM computers, an IBM 604 Electronic Calculating Punch and the IBM 650, to speed the processing of data in its Finance Division. But researchers soon expanded the machines' use to other things, including trajectory calculations. These were the early days of computers, and for all their speed, the machines made too many mistakes to earn the immediate trust of those who were risking their lives to advance the nation's space goals. But Dorothy Vaughan, the former head of the West Computers, had looked at the monstrous machines and caught a glimpse of the future. She'd taken classes in the evenings and on weekends and learned Fortran, the computer programming language that enabled her to feed the engineers' complex equations into the machine and translate the answers that it spit out. By 1960, Langley had consolidated its electronic computers and those who operated them into the Analysis and Computation Division to serve all of the research departments. Dorothy had landed there as a computer programmer, joining other Negro and white women who had worked as human computers.

Each generation of electronic computers grew more sophisticated, more powerful. By the end of 1960, the IBM 704 had been installed at Langley in the Analysis and Computation Division, and two IBM 7090 computers had been sent to a NASA facility in Washington, DC. The computers arrived just before the first scheduled launch of Project Mercury's orbital flight. But that date and several others would come and go as the engineers conducted more tests, fixed glitches, worked on the worldwide communications network, and did all they could to prevent a disastrous end. While they worked, the Soviets sent yet another of its cosmonauts, Gherman Titov, into space on October 6, 1961, for a record-breaking seventeen orbits. He spent an entire day in space, long enough to take photographs, experience a kind of motion sickness that caused him to vomit, and even fall asleep. As news of the Soviets' most recent exploits in space made it to the United States, NASA once again was at the center of public criticism and doubt about its ability to overcome the opponent's dominance in the skies.

Astronaut John Glenn, a former US Marines test pilot, had been the one anointed to carry out Project Mercury's ultimate goal of orbital flight. The nation's hopes were pinned on him also to resurrect the prospect that the United States could someday reign in the realm beyond Earth. And John Glenn had long been ready. He had hoped to be the first man in space, but instead he would become the first in orbit. He prepared himself meticulously, staying in top physical shape, submitting himself to regular monitoring by doctors, practicing his water exit from the capsule repeatedly, and participating in hundreds of simulated missions. After multiple delays, his date with the skies was all set: February 20, 1962.

John Glenn had been nearly obsessive about checking and rechecking every aspect of the flight that he could control. As he went through his last flight simulation, he had a final request of the engineers. He needed assurance that the trajectory plotted by the powerful IBM 7090 computer was correct.

"Get the girl to check the numbers," he instructed.

I was sitting at my desk when the phone call came in to one of the guys. He sat close enough for me to overhear his side of the conversation and figure out what was about to happen. As the engineer relayed the request to me, I remained calm. This was a major assignment, but I had done this long before the computer made it seem simple. So I quickly assembled my meager supplies and got busy on my calculator, working out every equation by hand for the trajectory of a mission that was scheduled to include three orbits. The computer had figured it out, but I was the error checker, the last stop. This was a meticulous task with no room for error. And digit after mind-exhausting digit, I computed, filling a thick pile of data sheets. Every step of the way, I paused to check my numbers against the computer's to make sure there was agreement. I even carried out the calculations a couple of decimal points beyond the computer's numbers. One and a half days later, I finally finished. I took a long, deep breath. The numbers agreed. Our astronaut was ready to go.

The space team had decided to use the newer, faster Atlas rocket to boost Glenn's *Friendship 7* flight from Cape Canaveral into space (a decision that the *Hidden Figures* movie attached to me, I suppose to add to the drama). The launch, initially scheduled for December 1961, was delayed four times because of mechanical issues or poor weather conditions. Then finally, just before 9:47 a.m. EST on February 20, a live television audience of about 135 million people watched and listened as Mission Control performed its final system checks.

"May the good Lord ride all the way," test conductor Tom O'Malley said.

Scott Carpenter, who had been selected as the backup astronaut for the flight, added, "Godspeed, John Glenn."

With that, *Friendship* 7, carrying America's astronaut, was blasted into the heavens. I watched with butterflies in my stomach from a television in the office. The takeoff looked perfect. I waited, worked, and waited some more. Then, after four hours, fifty-five minutes, twenty-three seconds, and three orbits around the Earth, the broadcast returned to the ocean splash-down, which looked perfect, too. The capsule landed just about forty miles off target, and I later learned that was because it carried the unanticipated added weight of the rocket, which had stayed in place. Near the end of the second orbit, a capsule alert had indicated that the spacecraft's heat shield was loose. Without it, the capsule and its passenger quickly would have succumbed to the overwhelming heat, about three thousand degrees Fahrenheit, as they headed back down to Earth. But Mission Control had made an on-the-spot decision, instructing Glenn not to jettison the rocket, as planned, with the hope that it would keep the heat shield in place. It may have been a life-saving decision. Glenn also had overcome a problem with the automatic control system by switching to manual and operating it himself.

After the successful mission, Americans couldn't get enough of our newest space hero. Tens of thousands of Hampton and

Newport News residents lined the streets a few weeks later to celebrate Glenn, who rode in the first of a fifty-five-vehicle motorcade that also included the six other astronauts and their families. Leaving from Langley, the parade traveled twenty-two miles, passing all of the familiar landmarks, including the shipyard and Hampton Institute. I watched with some of the engineers at Langley. The parade ended at Darling Stadium, where another jubilant crowd was waiting.

John Glenn had caught us up to the Soviets. Now the Space Race seemed ours to win.

Many have asked me over the years whether John Glenn ever knew my name. Who knows? It didn't matter to me then, and it doesn't now. It was enough for me that *I knew* when he needed "the girl" to boost his confidence that he could entrust his life to the heavens and get him back home, I was that girl. I loved my job more than ever. And I felt blessed to be her.

chapter 9

SHOOT FOR THE MOON

Somehow, word got out in the Negro press that I had played a role in John Glenn's historic flight. Back then, Negro newspapers and magazines, especially *Ebony* and *Jet* were how we kept up with what was going on with our people. The other media acted as if we didn't exist, except for crime stories. Practically every Negro home had copies of *Ebony* on the coffee table. The magazine showed us at our best in all colors, from the brightest beige to the deepest chocolate. The stories and photos gave us a glimpse of the high life of our dreams. Some ladies wore their hair, chose their outfits, and decorated their homes like something they saw in *Ebony*. Negro newspapers, including the *Chicago Defender,* the *Afro-American,* the *Pittsburgh Courier*, and the *Norfolk Journal and Guide* spoke our truth. They told us the horror of Emmett Till's murder and sent reporters to Little Rock Central High School and the student sit-ins in North Carolina long before the white media figured out they were stories worth covering. They questioned why Negro soldiers were fighting for freedom abroad for a country that treated them as second-class citizens at home. They wrote about our marriages, our deaths, and our sons and daughters graduating from college or receiving a military honor. So when a reporter called me from the *Pittsburgh Courier* sometime after John Glenn's flight to interview me for a story, I was at once a bit surprised, nervous, and somewhat confused. I didn't talk much about my job outside of the office, and it was the first time that my work was being recognized as something special. It was work to me, but perhaps a story could inspire a young person to consider a career as a mathematician. So I agreed to the interview.

When I saw the story on March 10, 1962, in the national edition of the *Pittsburgh Courier*, I was shocked. There was a big photo of me stretched three columns across the front page. The headline read, "Lady Mathematician Played Key Role in Glenn Space Flight," and the story on an inside page called me "one of the most brilliant mathematicians of the present era." *Oh my*, I thought. This was too much. But it was a very positive story, and I liked that it mentioned Joylette was editor in chief of the *Hampton Script* newspaper in her senior year at the Institute. I wish it had mentioned my other two talented daughters as well.

Just beneath the story, another headline asked, "Why No Negro Astronauts?" The article actually focused on the comments of a school desegregation attorney named Paul B. Zuber, who criticized the federal government for failing to select a Negro in the pool of astronauts. NASA responded that the selection of astronauts had nothing to do with race but with qualifications and that there would be Negro astronauts in the future when more Negroes entered the science fields. I looked forward to that day. Times were changing, and the next generation of Negroes would be able to aim higher than they could see, maybe even as high as the moon.

I was seeing evidence of the changing times in my own family as Joylette prepared to enter the work world. Before graduating from Hampton Institute, she had decided that she didn't want to follow the traditional path of college-educated Negroes into the classroom to teach. In my generation, most Negroes with degrees in subject areas such as math and science had few other options but to teach. But in the spring of her senior year of college, Joylette applied to NASA for a job as a mathematician. Then a flurry of major life events happened for her at once. She graduated from college on June 6, 1962, and sixteen days later eloped with her college sweetheart, Lawrence Hylick. Lawrence had asked Jim and me for Joylette's hand in marriage months earlier, on Christmas. We were happy for them and agreed wholeheartedly. But the couple told no one about their wedding plans and held a secret ceremony in the chapel on Hampton Institute's campus, where they'd met.

Lawrence, who had another year to complete at the Institute, was set to travel out of town for a couple of months to ROTC summer camp, and the two of them decided to get married before he left. The news was quite a surprise, but they were two young adults in love and in a hurry to start their lives together. I certainly knew something about that. At about the same time, Joylette got more good news—she had landed the job at NASA. She joined me at Langley just four days after her secret wedding. My oldest daughter was now a mathematician, like me. She worked at NASA, like me. And she had started right out of college. That showed me how important it is for young people to be able to see in the flesh a vision of what they could become. But it also showed me how critical it is for those with influence—parents, teachers, and mentors—to help those who look to them for guidance to envision for themselves what may seem impossible. Dr. Claytor's vision for me had in essence given birth to two generations of mathematicians. He had opened my mind and my dreams to a job I never even knew existed, and by watching me go to work at Langley every day, my daughter knew she didn't have to become a teacher if that wasn't her heart's desire. Now, that made me feel proud.

When Lawrence returned from summer camp, Jim and I planned to host a reception for the newlyweds in the backyard of our house on Mimosa Crescent. But early on the morning of the planned celebration, August 3, 1962, a relative called with tragic news. My nephew, Lieutenant Eric Epps Jr., the oldest son of Margaret (my deceased husband Jimmie's sister), and Eric Epps Sr. had been killed in a military plane crash. Margaret and Eric Sr. were the couple who had invited Jimmie, the girls, and me to Newport News all those years ago. The oldest of three boys, Eric Jr.—"Skippy" to us—was so bashful that when Jimmie, the girls, and I walked into the back door of their home for the first time, he and his two brothers dashed out the front door. But the Epps boys got to know their cousins, and the children soon were inseparable. Skippy always loved planes, and his parents nurtured the notion that a smart, curious Negro boy could grow up and fly planes someday. Model planes hung from the ceiling in his bedroom. This was in the early 1950s, just about a decade after Tuskegee Institute produced its first class of Negro pilots through the federal government's Civilian Pilot Training Program. The federal government had financed the program at colleges and flight schools throughout the country in a desperate attempt to train more pilots for what appeared to be an inevitable third world war. HBCUs were not initially included because the prevailing attitude was that our people were not intelligent enough to learn how to fly planes. But after criticism from Negro newspapers and activists, the federal government conceded and included select HBCUs in the program. I've always been proud that my alma mater, West Virginia State, was among the first six schools selected in 1939 to start an aviation training program. It was led by the university's vocational and technical director, Mr. James Evans, who had played such an important role in my life. Tuskegee was selected, too, and as its program quickly expanded to train commercial pilots and then officers for the Army Air Corps, West Virginia State and the other schools became feeder schools into Tuskegee's airman training program. Two graduates of West Virginia State's program—George "Spanky" Roberts and Mac Ross—were among the first five to complete Tuskegee's program and become fighter pilots and commanding officers in the US Army. The Negro community, and especially those of us connected to West Virginia State and Tuskegee, were so proud of the airmen. Tuskegee trained an estimated eleven hundred combat pilots, who would go on to serve with unmatched distinction in World War II and change minds across the nation about the intelligence and academic abilities of Negroes.

My family watched Skippy grow into a fine young man who was our shining star. He graduated from Huntington High School in 1956 and attended the University of Michigan, where he majored in aeronautical engineering. After two years, he enrolled in the cadet training program at Lackland Air Force Base in San Antonio, Texas. By age twenty-three he had risen to the rank of first lieutenant in the US Air Force, and he was the first Negro pilot at Pease Air Force Base in New Hampshire to fly a B-47 jet bomber. Eric Jr. quickly became a rising star there, too, and had been selected as part of a topflight crew to represent the 509th Bomb Wing in the Strategic Air Command bombing competition the following month. He was copilot of the plane, which went down shortly after taking off from the base during a training mission. My nephew was just twenty-three years old—ten days shy of turning twenty-four (though all the newspapers reported that he was twenty-four). Two others also were killed in the crash: Captain Eugene S. Procknal, the thirty-two-year-old pilot, a married father of four, whose oldest child was just four years old; and First Lieutenant Edward R. Sowinski, twenty-five, the navigator, who left behind a young wife. Those two men actually lived with their families on the air force base where they died. All three seemed so full of promise. Witnesses reported hearing the jet take off and what sounded like the engine shut off just before the fiery crash.

The news crushed our family. But after the tears, hugs, and words of comfort to one another, we decided we should still move forward with the reception for Joylette and Lawrence. Lawrence's parents were on their way from New York, and other out-of-town relatives were on the road as well. Our hearts were heavy, but we painted on smiles and did our best to celebrate our family's newest couple.

Lawrence returned to Hampton in the fall, and Joylette continued to enjoy her job in the Reentry Physics Branch, where she worked mathematical equations from the engineers on data sheets and transferred them to a large stack of data cards. The cards were then read into the computer. Joylette worked in a different building, near the wind tunnels, and we never saw one another at work. We rarely even talked about work at home. But I was proud just knowing she was there.

The next year, in early summer 1963, Kathy returned home to Newport News to transfer to Hampton Institute. She had planned to attend Bennett for just two years, and the experience gave her an opportunity finally as the youngest child to stretch out on her own and figure out who she was without her older sisters. Connie got married to a Hampton graduate named John Boykin, and they moved to New Jersey. My girls had grown into active, independent young women, and as they moved onto their individual paths, I wondered what kind of world would emerge around them.

The summer of 1963 was such an uncertain time as racial protests intensified in the South, forcing us as a nation to decide who we wanted to be. Were we really a nation that could look away from the protests in Birmingham as police officers released their dogs on Negro children, beat the children with billy clubs, and as firemen turned hoses on them with a force that ripped off shirts and pushed bodies down the street? Those were the disturbing images that drew me into the events in Birmingham in early May 1963. There had been earlier civil rights marches in that city in defiance of a circuit judge's ruling that had made such protests illegal. Dr. Martin Luther King Jr., fellow organizers Ralph Abernathy and Rev. Fred Shuttlesworth, as well as other protesters had been arrested on Good Friday, April 12, 1963. But the national media hadn't given the protests or the arrests significant attention. Dr. King's "Letter from a Birmingham Jail," written in response to white ministers who criticized his presence in Birmingham, explained powerfully why Negroes could no longer wait for justice. But when I clicked on my television that first week in May and saw the attacks on children, I couldn't believe my eyes. I couldn't believe the inhumanity of it. And as a mother, I could only imagine the terror, anguish, and anger their parents must have felt. I had disagreed with my own college students' participation in the sit-ins, so I didn't like the idea that even younger children's lives had been put at risk during the Birmingham protests. But I'm sure no one could have imagined the horror those young people would face at the hands of the heartless police commissioner Eugene "Bull" Connor, who had ordered such tactics. My heart bled for those children and their parents, and I couldn't take my eyes off my television screen. Neither could the rest of the world. And, as it turned out, that made all the difference.

A bit of history here . . . the Birmingham Children's Crusade, as the demonstrations would come to be called, had been organized by one of the movement's young stars, James Bevel, to bring attention to the struggle in one of the nation's most staunchly segregated cities. Movement organizers debated whether to use children for the protests, but when Dr. King agreed, Bevel, and other young organizers came to town and began recruiting and training the youths in nonviolent resistance. Negro deejays helped spread the word, and on May 2, 1963, the day they were calling D-Day, hundreds of students walked away from their schools and made their way over to 16th Street Baptist Church. From there, the students marched out of the church, two by two, in groups of fifty at a time, singing freedom songs and moving across the street toward Kelly Ingram Park. The police arrested them, and one group kept replacing the other until about a thousand young people had filled the Birmingham jail to capacity. When just as many students returned to protest the second day, Connor tried a stronger deterrent: the vicious dogs and fire hoses. By then, the white news crews were watching and broadcast Connor's shameful tactics for the world to see. Public outrage beyond Birmingham grew. The international embarrassment and public pressure pushed President John F. Kennedy and his brother Attorney General Robert Kennedy to use their influence to nudge the white establishment in Birmingham to negotiate with the protest leaders. By May 10, Dr. King and Rev. Shuttlesworth announced a major victory with an agreement by local officials to desegregate public facilities, end discriminatory hiring practices over a period of ninety days, and to continue meeting with Negro leaders. The community had hardly been able to grasp the announcement fully when the next day a bomb ripped through the Gaston Motel, the Negro-owned business where Dr. King and other out-of-town movement leaders had been staying. Fortunately, they had left just hours earlier. The home of Dr. King's brother A. D. King was also bombed. There were no major injuries or deaths in the bombings, but the battle for freedom was heating up, and the segregationists were making it clear that they were not conceding any ground without a fight.

It was against this solemn backdrop that the US Department of Labor published a brochure recognizing the centennial of the Emancipation Proclamation in May 1963. It featured a photo of a little Negro boy wearing no shoes on the cover and comments inside from President Kennedy and Vice President Lyndon B. Johnson about the strides Negroes had made in this country since the end of slavery. A photo of me at my desk was included, along with seven other men who also worked in the space program at NASA. The title of the brochure announced: "America Is for Everybody." My parents certainly had raised me to believe that, but hateful forces were battling for a different kind of America.

About a month after the brochure's hopeful declaration, avowed segregationist governor George Wallace blocked the entrance to the University of Alabama's auditorium as two Negro students attempted to complete their registration and desegregate the campus, which had been cleared by a federal court. The Kennedy brothers again were forced to intervene, sending a hundred troops to protect the two students. The same day, President Kennedy appeared in a televised speech to talk about the action he had taken in Alabama and explain to the American people why he had proposed perhaps the most comprehensive civil rights bill ever. I'd never heard a white man talk so eloquently about race as when he gave that speech.

The same evening in Jackson, Mississippi, Myrlie Evers and her children had watched President Kennedy's speech and were waiting for her husband, Medgar, to return home from an NAACP meeting about a half hour after midnight when they heard gunshots. Myrlie rushed outside and found her thirty-seven-year-old husband shot in the back and near death on their front steps. He had been shot as he walked from his car to his home, and he staggered up to the steps before collapsing. Medgar Evers, who had been organizing protests and voter registration drives as the NAACP's first field secretary in Mississippi, died within an hour. Our people were being targeted, intimidated, and killed across the South for demanding to live as equal citizens. And no matter where the rest of us happened to live in this country, every Negro felt the pain and frustration. That is surely what drove tens of thousands of Negroes and white supporters to the nation's capital on August 28, 1963, for what was being called the March on Washington for Jobs and Freedom. The civil rights agenda for the march had been set primarily by the leaders of six prominent civil rights organizations, the "Big Six"—Roy Wilkins, executive director of the NAACP; Whitney Young, executive director of the National Urban League; Rev. Martin Luther King Jr., chairman of the Southern Christian Leadership Conference; John Lewis, president of the Student Nonviolent Coordinating Committee (SNCC); James Farmer, founder of the Congress of Racial Equality (CORE); and A. Philip Randolph, organizer of the Brotherhood of Sleeping Car Porters.

It would become public in the years ahead that civil rights leader Anna Arnold Hedgeman, who was on the march's organizing committee, and Dorothy Height, head of the National Council of Negro Women, had to push to get even a puny recognition for the movement's female leaders onto the program. Just one of them—Daisy Bates, the fireball who had advised the Little Rock Nine in Arkansas—got to say a few words. And even then, she had to pledge the women's support for the men. The names of a few other women were called: Gloria Richardson, president of the Cambridge Nonviolent Action Committee (CNAC), who led protests in Massachusetts; Rosa Parks, whose civil rights activism extended far beyond the Montgomery bus boycott; Diane Nash, a key organizer of the student sit-ins in Nashville in 1960 and a critical figure in many of the major protests afterward; Myrlie Evers, the widow of Medgar Evers, who had worked by his side and picked up his crusade upon his death; and Prince Lee, whose husband was murdered in Liberty, Mississippi, for his involvement with SNCC.

I never marched like those brave women, but my heart was always with them and the movement. I'm a proud lifetime member of the NAACP, which was at the forefront of many of these important civil rights battles. I'd like to think that I did my part from the inside. Every time I did my job to the best of my ability and succeeded, just maybe I was proving to those who would discount women and Negroes that we were just as capable as anyone else—if not more. And every time I pushed to go where no woman or Negro had gone before and won, it was a victory for us all.

Despite the celebrities who spoke and sang at the March on Washington that day, ordinary people made the event historic. An estimated 250,000 to 300,000 of them flowed from every crevice of the country into the District of Columbia and onto the National Mall—the largest demonstration for civil rights in the country at that point. As the people stood there, Negroes and whites together under the blazing sun, no one knew whether the moment would spark more violence or change. Lord knows, there had been too much violence already: the attacks on Freedom Riders who had been beaten in South Carolina and Alabama and their bus firebombed in May 1961 as they traveled south to force the desegregation of interstate travel; the bloody riots that broke out when James Meredith, a Negro US Air Force veteran, tried to exercise his right to register at the segregated, all-white University of Mississippi a year later; Bull Connor's wickedness; Medgar Evers's murder; and all the hate-inspired violence that never made it onto the media.

Yet the people would not be stilled by fear. They marched on Washington and filled every space between the Washington Monument and Lincoln Memorial with their hopes. Dr. King grabbed hold of that hope and electrified the crowd—those under the sound of his voice and those of us who watched later on television—with his now famous "I Have a Dream" speech. His dream and ours would require America to change. As it turned out, change was coming, but so, too, was more violence.

On September 15, less than a month after that historic march, the 16th Avenue Baptist Church in Birmingham, the meeting place for many of the protests, was bombed. And four precious girls—Addie Mae Collins, Denise McNair, Carole Robertson, and Cynthia Wesley—were killed in the blast. I could not fathom the depth of hatred that would set in motion such an evil scheme. And I was haunted by the smiling faces of those pretty little girls. I thought hatred couldn't get any worse than that. Then, on Friday, November 22, I got a call at work that I will never forget. It was Joylette, sobbing.

"President Kennedy has been shot," she said between her cries. He was dead.

Joylette was home from work that day with her new baby girl, Laurie, my first grandchild. The news seemed to suck the air from the room and drain the blood from my face as I sat there, speechless. Someone had murdered the President of the United States? In my heart I knew why. The Kennedy brothers had been allies in our people's fight for freedom. About 70 percent of Negroes had voted for President Kennedy in his 1960 presidential election, including Jim and me. I remembered the time when many of the Negroes I had known in West Virginia were Republicans, the party of Abraham Lincoln. But President Kennedy and his brother had intervened to protect civil rights leaders and protesters in some of the hardest-fought battles, and in the process they won Negro hearts. Now President Kennedy was gone. For a long while, the world just didn't feel safe.

The following January, my hope that Joylette would enjoy a long career at NASA burst when a frustrating job search for her husband pushed the two of them to relocate to New Jersey. Lawrence had graduated with a degree in accounting in June 1963, but after months of searching for a job in the Hampton Roads area, he got no offers. He sought assistance from the state employment agency but was told there were no jobs for Negro accountants. Lawrence decided to return to his native New York City, where there were more opportunities for Negro professionals. Even there, he had some disappointing racial encounters, like never hearing back from a company after scoring high on a required test. Lawrence ultimately landed an accounting position at a company in New Jersey through a listing in the *Amsterdam News*, a Negro newspaper based in Harlem. The bubble around me at Langley may have been just a bit more progressive, but Virginia was still the South, and it was upsetting that my college-educated son-in-law had to move north with my daughter and granddaughter in January 1964 to pursue the career of his dreams.

The following summer, the civil rights struggle seemed to reach a crescendo in the Deep South as Negro leaders pushed forward with massive registration drives, only to experience more violence. Young adults from all over the country were recruited to travel to Mississippi to participate in a major campaign, called the Mississippi Summer Project, or Freedom Summer, to increase voter registration among Negroes and provide literacy and other training in "freedom schools." About seven hundred volunteers joined the Mississippi crusade, and there was trouble right away. When I heard on the news in June that three of the young volunteers were missing, I suspected the worst. Never had I been sadder to be right than when I heard the report two months later that investigators had found the dead bodies of twenty-one-year-old John Chaney, a Negro from Meridian, Mississippi; twenty-year-old Andrew Goodman and twenty-four-year-old Michael Schwerner, both of Jewish heritage from New York City. Goodman and Schwerner had been shot once in the heart, and Chaney had been beaten brutally before he, too, was fatally shot. They were buried together in an earthen dam just outside Philadelphia, Mississippi, on a farm owned by one of the white supremacists involved in the murder. President Lyndon B. Johnson, the former vice president who had assumed the presidency after President Kennedy's assassination, had dispatched the FBI to help find the missing volunteers and even the US Navy to dredge local swamps and waterways. While doing so, the investigators found the bodies of eight other Negroes, including two college students and a fourteen-year-old boy, who all had disappeared mysteriously, as well as five others who were never identified. At the time, few people outside of Mississippi even heard about those happenchance discoveries because the national media barely even noted them. Why? Because they were not the two white civil rights volunteers. They were not the ones who really mattered then to an outraged public. That, too, was tragic. I lived far away from Mississippi, but the murders of Chaney, Goodman, and Schwerner sent a chill up my spine. These young men were about the ages of my girls and probably had been full of the same idealism and hope. Their horrific end was the nightmare of every conflicted parent whose child had even thought about fighting for needed change in the trenches.

The three freedom workers had yet to be found when President Johnson pushed forward with the civil rights legislation that had been proposed by his predecessor. The news reports and even a meeting with the parents of Goodman and Schwerner likely helped build support for the measure. Southern Democrats tried unsuccessfully to stop it with an historic filibuster, but Congress approved the groundbreaking legislation. Called the Civil Rights Act of 1964, it aimed to protect Negroes against the voter qualification tests that had routinely disqualified so many of us from voting; it outlawed discrimination in hotels, motels, restaurants, theaters, and other public businesses involved in interstate commerce; it authorized the US attorney general to file lawsuits to force the desegregation of public schools; it authorized the withdrawal of federal funds from programs that practice discrimination; it banned discrimination in employment in businesses with more than twenty-five people; and it created the Equal Employment Opportunity Commission to review discrimination complaints. President Johnson signed the measure into law on July 2, 1964. I was hopeful that after so much turmoil the country would settle into a more peaceful and just place.

But the racial strife continued. Election officials in the South still found ways to deny Negroes the right to vote, which prompted more protests. On March 7, 1965, hundreds of protesters attempted to march from Selma, Alabma, to Montgomery, the state capital, but as they headed over the Edmund Pettus Bridge, they faced a wall of local deputies and state troopers, including some on horseback. The officers moved forward, attacking the marchers with tear gas, billy clubs, and electric cattle prods. One of the young leaders, then twenty-five-year-old John Lewis, who eventually would become a longtime congressman, was beaten so badly that his skull was fractured. But television broadcasts of the attack, known thereafter as "Bloody Sunday," pushed President Johnson a week later to announce in a nationally televised speech that he was sending additional civil rights legislation to Congress.

"There is no Negro problem. There is no southern problem. There is no northern problem. There is only an American problem," President Johnson said in the speech. "And we are met here tonight as Americans—not as Democrats or Republicans—we are met here as Americans to solve that problem."

After congressional approval, President Johnson signed the Voting Rights Act of 1965 into law on August 6, 1965. By then, the United States was engaged in a controversial war in Vietnam. And once again, the country began drafting young men to fight overseas. At about this time, the whole country seemed to take a chaotic turn. Young white activists primarily were protesting the Vietnam War, and young Negroes were parting ways with Dr. King and the nonviolent Civil Rights Movement for a more radical approach. They began growing their natural hair into Afros (and in my girls' case, trying to curl it tight to resemble an Afro), and "black" soon replaced "Negroes" and "colored" as the preferred term for our race. The "Black Power" message of racial pride, self-love, and economic empowerment was appealing. Those values had been ingrained in me from childhood. But I felt out of step with the changing styles and more combative rhetoric.

My adult children were living on their own, embracing our evolving culture, sporting their Afros, and wearing African-inspired head wraps and clothing. Lawrence and Joylette had settled in East Orange, New Jersey, and she became a homemaker for several years to take care of Laurie and later a son, Troy. Lawrence eventually enrolled at Columbia University and earned an MBA degree. Connie and John, also in New Jersey, eventually added three children to their family—Michele, Greg, and Doug. Kathy had married her college sweetheart, Donald Moore, in a July 1965 ceremony at Carver Memorial, and they were living in Texas. He had graduated from Hampton as an architecture major and second lieutenant from the army's ROTC there. We worried that he would be deployed to Vietnam, but fortunately he was stationed in El Paso in a teaching role. With our nest empty, Jim and I enjoyed a busy social calendar, and I continued doing what I had always done in good and bad times: serving the community through my church and sorority and working harder than ever.

The pressure was intense at work to meet President Kennedy's goal of a moon landing by the end of the decade, though he did not live to see us get there. The space project designed to follow Mercury had been named already during the administration of President Dwight Eisenhower. It was called Apollo, reminiscent of the Greek god who rode his mythical gold chariot across the sky to bring light to the world. Apollo had been envisioned to carry three astronauts on a variety of missions that included an orbit of the moon. With President Kennedy's pronouncement, Apollo was dedicated to landing on the moon and doing so under an advanced timetable calculated to beat the Soviets. At the time we were still behind in the Space Race, and our astronauts had yet even to orbit the Earth. There was still so much to learn before conquering the moon. NASA engineers soon recognized that they would need a bridge project to develop and test the technology to accomplish their ultimate goal. That bridge was christened Project Gemini, Latin for "twins," which would carry two astronauts and use two vehicles.

While Project Gemini was under development, NASA's first space venture, Project Mercury, wrapped up its missions. It had accomplished its goal of determining that a human could survive in space. On May 16, 1963, astronaut Gordon Cooper flew on Project Mercury's final mission with a twenty-two-orbit trip over a period of more than thirty-four hours—the longest time yet for one of our astronauts in space. The next year, the first unmanned Gemini test flights were launched, followed in 1965 and 1966 by ten flights with crews that laid the groundwork for Apollo. Gemini astronauts performed a number of important missions that involved changing orbits, rendezvousing with other spacecraft, docking maneuvers, walking in space for the first time, and spending long periods of time there to evaluate the effects. It was also determined that a larger Mission Control Center, which directs each space flight and provides around-the-clock monitoring and technical support, would be needed to send astronauts to the moon. So the Mission Control Center operations moved from Cape Canaveral, Florida, to the Manned Spacecraft Center in Houston with the *Gemini 4* mission in 1965.

All the while, I worked closely at Langley with engineer Al Hamer on trajectory calculations for the trip to the moon and contingency plans in case of disaster, including how to navigate the spacecraft back home using the stars as a guide if there were an electrical failure. Al and I examined these and other technical topics in four papers that we would publish together between 1963 and 1969. He and I also became great friends. We played bridge at lunchtime with John Young and other engineers in our department. When Al married his wife, Virginia, he planned to invite me to his wedding, but he came to me one day with a pained expression on his face. He said he had gone to his pastor in advance of the wedding to let him know that some friends of his, who happened to be Negro, would be attending. He didn't want there to be any trouble. He was surprised and disappointed when the pastor told him that Negro guests would not be allowed to attend. Al was so apologetic and angry when he told me, but I just shrugged it off. I was not going to allow his pastor's backward views to change my opinion of the lovely couple. We would socialize in one another's homes many times in the years to come.

Finally, NASA was ready to launch the first Apollo flight. On January 27, 1967, less than a month before their scheduled launch, astronauts Virgil I. "Gus" Grissom (one of the original seven astronauts in Project Mercury), Edward H. White II, and Roger E. Chaffee climbed aboard the spacecraft for a launch countdown rehearsal at the site in Cape Canaveral. The command module spacecraft was attached to the powerful Saturn 1B rocket, but it was not fueled. Several minor problems caused significant delays, and then at 5:40 p.m., more than four and a half hours after the astronauts boarded, the communications system failed. At 6:30 p.m. Grissom remarked, "How are we going to get to the moon if we can't talk between three buildings?" A minute later something went terribly wrong, and a fire broke out in the cabin. The flames and toxic smoke quickly intensified and overwhelmed the three astronauts, who were unable to open the inward-opening emergency hatch because of the cabin's internal pressure and several safety latches. Thick black smoke and toxic fumes flowed heavily from the aircraft and kept pushing nearby technicians back. It was later determined that their gas masks were ineffective. By the time workers outside the cabin were able to remove the hatch and get to the astronauts, all three were dead.

It was late on that Friday evening when I heard the devastating news. This was an unspeakable tragedy that deeply shook everyone even loosely connected to the space program. It made the risks that these space explorers were taking with each journey—and in this case, even preparing for the journey—more real than ever. It was even more difficult, I think, because this catastrophe occurred on Earth, on the ground, in full view of everyone onsite. Yet no one was able to save our guys.

The main cause of the fire was determined to be electrical, but because the rocket wasn't fueled, the flight test wasn't deemed hazardous, and emergency preparedness had not been adequate. These were among the findings of an exhaustive investigation into the cause of the fire and the astronauts' deaths. There were also congressional investigations and hearings that questioned NASA's leadership. While none of us employees knew what the political fallout would be or what the future would hold for the moon mission, we just kept working. We all seemed more determined than ever to do our parts, to learn from any mistakes, to work harder and smarter, and to do all within our power to avoid this kind of devastation again. The spacecraft was redesigned, and every weakness that was identified was addressed.

We had shot for the moon and watched the dream go up in flames. But while the destruction was great, the dream itself was not consumed. As in life, we at NASA learned from the remnants of the *Apollo 1* tragedy. We grew stronger, wiser, and kept our sights on the moon.

chapter 10

FINISH STRONG

Just eleven months after the three *Apollo 1* astronauts died in that tragic fire, another space program tragedy consumed the life of the country's first black astronaut. But the fatal supersonic jet crash that killed Major Robert H. Lawrence Jr. on December 8, 1967, at Edwards Air Force Base in California, went largely unnoticed. Few people, including me, had even heard his name. Three more decades would pass before he received the proper recognition.

Major Lawrence was among seventeen air force pilots chosen for a space venture, the Manned Orbiting Laboratory (MOL), a joint project of the US Air Force and the National Reconnaissance Office, according to multiple published reports. The program's secret mission was to launch crews into space and use mini-space stations to obtain high-resolution photographs of our country's Cold War enemies. But Major Lawrence never got to carry out that mission. He was the backseat instructor in an F-104 *Starfighter* supersonic jet with another pilot, who was practicing a landing technique—one that would be used in future space shuttle landings—when the jet crashed. Both pilots ejected, but Lawrence, then age thirty-two, was killed instantly. Because of the secrecy of the MOL project, Lawrence's role was not widely publicized. But controversy erupted in May 1991 when the Astronauts Memorial Foundation dedicated its new Space Mirror Memorial at the Kennedy Space Center in Cape Canaveral to sixteen astronauts who had died in the line of duty, and Lawrence's name was not included. Under the air force's rules, Lawrence was not considered an astronaut because he had not completed his training and had not flown at least fifty miles above the Earth's surface. At NASA, however, a person is considered an astronaut upon acceptance into its astronaut training program. So the name of *Apollo 1* astronaut Roger Chaffee and other NASA astronauts killed before making it to space were included on the memorial. It took years of lobbying by Lawrence's widow, Barbara; his son, Tracey; other family members; and political supporters before the decision-makers did the right thing and finally added Lawrence's name on the memorial in 1997, thirty years after he died. The MOL program that had trained Major Lawrence for space was canceled in 1969, and its seven youngest astronauts, Major Lawrence's contemporaries, were transferred to NASA's space program.

I'm often asked if I knew Major Lawrence, but I can only imagine how much I would have enjoyed meeting such a brilliant young man, who not only had graduated from high school at age sixteen but also earned a PhD in physical chemistry from Ohio State University in 1965 and was a classical pianist. I can only imagine the bridge he could have been to connect NASA to a skeptical Black community that in 1969 felt excluded and left behind as the space agency explored the heavens on the way to the moon.

Many in the black community had never forgiven NASA for the mistreatment of Ed Dwight, who in the early 1960s seemed on track to become the first black astronaut. Dwight was a promising US Air Force pilot with an aeronautical engineering degree from Arizona State University when the Kennedy administration chose him in 1962 to enter the test pilot training program at Edwards Air Force Base in California. NASA selected its astronauts from the training program, and President Kennedy had been eager to find a black pilot who met the qualifications for selection. Black voters had helped to catapult Kennedy into office, and a black astronaut surely would provide an inspiring symbol of hope during tense racial times.

I never met Dwight during those years, but his selection and ultimate completion of the program quickly caught the attention of the black press, and his picture appeared on the covers of the major Black magazines, with accompanying stories touting him as the one most likely to break the color line in space. But when NASA announced its next group of astronauts in October 1963, the ones who most likely would explore the moon, Dwight was not among them. The introduction of the fourteen members of Astronaut Group 3—all white men—punctured the inflated hope of an entire community, which felt betrayed. The June 1965 issue of *Ebony* magazine featured a long, investigative story that raised questions about Dwight's experience in the training program. "Still unavailable is a complete accounting from the military-space bureaucracy of the reasons for the apparent stunting of Dwight's career in space before it ever actually began," the story said. "Was Dwight rejected by the National Aeronautics and Space Administration (NASA) for additional astronaut training at its big manned spaceflight center in Houston for purely technical reasons? Or did other factors—such as Dwight's race—enter into the decision to deny him a possible role in NASA's earth-orbiting Project Gemini or the moon venture, Project Apollo?"

Frustrated, Dwight ultimately resigned from the air force in 1966. He later discussed in interviews and in his autobiography the racism he had experienced during the training program at Edwards Air Force Base. He said he was derisively called "Kennedy's boy" and that when President Kennedy was assassinated a month after the Astronaut 3 group was announced, any chance he might have had to become an astronaut died, too. But despite that searing pain and disappointment, Dwight would go on to pursue his first love, art, and he ultimately became an internationally acclaimed sculptor of African American history. He became a shining example of how to persevere and finish strong when life dishes out devastation beyond your control. I was honored in 2016 to receive an award that was sculpted by Dwight and named in his honor from a black aviators' nonprofit organization, Shades of Blue.

By working inside the bubble of NASA and rarely talking about my work outside of the office, I had little understanding of the wide gulf between my employer and my people in the late 1960s, as NASA inched closer to the moon. But that tension soon would bubble to the surface for all the world to see at the most critical moment. First, though, the country would suffer another devastating loss. On April 4, 1968, Dr. Martin Luther King Jr. was shot to death on the balcony of the Lorraine Motel in Memphis, Tennessee. His assassination left a hole in the heart of the Civil Rights Movement, and to those of us who believed in the ideals he espoused so eloquently, his death felt personal and gut-wrenching. It didn't make sense that a man who had spent so much of his life advocating for peace and justice died so violently. But in times like this, I knew better than to lean to my own understanding. I sought solace in my faith. Jim and I grieved with our church family at Carver Presbyterian. Kathy called us from Hampton Institute and told us that the chapel bell on campus was ringing solemnly. She and other students on campus were gathering in their dorm rooms to mourn together and comfort one another. Jim and I prayed for the safety of Joylette and Connie and their families over the next few days as television news reports showed riots spreading to Washington, DC, Baltimore, New York City, Chicago, and other major cities in response to Dr. King's murder. Enraged rioters set businesses on fire, broke windows, looted from stores, clashed with police, and many lives were lost. We were all hurting, but the retaliation and destruction were so opposite to everything Dr. King and the Civil Rights Movement had represented. I wondered when these troubled times would end.

But things only got worse. Just two months later, Robert F. Kennedy, by then a US senator, was shot to death at the Ambassador Hotel in Los Angeles. He had just won two important victories in the primary elections and seemed on his way to becoming the Democratic nominee in the upcoming presidential election. Our country seemed to be dissolving into chaos. Our soldiers were dying in record numbers in Vietnam; students on college campuses across the country were protesting the United States' involvement in the war; and younger, angrier black voices for change were drowning out the nonviolent Civil Rights Movement.

All I knew to do was burrow into my work. Ironically, the work that I produced during this time would make me prouder than anything else in my entire career at NASA. With the Gemini flights behind us, the Apollo crew was finally ready to get started. My job was in general the same as it had been in previous missions: to compute the orbit to the desired destination and back and compute the launch window, telling the engineers what time the astronauts would get to their destination and what time they would get back to Earth. The calculations for the moon orbit were quite intricate because there were so many factors involved. We had to consider, for example, the rotation of the moon—where it was at that moment, the number of days it would take the astronauts to get there, how far the moon has moved in that time, and where the moon would be by the time the astronauts made it there. The process reminded me a bit of what I was once told about hunting rabbits. A good hunter doesn't shoot right at the rabbit. You aim where you think the rabbit will be by the time your bullet gets there. We had to know where the moon would be by the time the astronauts arrived. Then, after computing the trajectory to get to the moon, I had to reverse the calculations to get them home. This time I had to consider how far the Earth has rotated. My partner at work, Al Hamer, was an expert in guidance and navigation. One of the interesting studies I worked on with him determined the amount of error that could be tolerated in an orbit. It was called the "error ellipsoid," which let the astronauts know at what point they would have to make corrections if they were not on the nominal trajectory. That was Al Hamer's baby.

After the *Apollo 1* tragedy, NASA canceled the planned follow-up mission, *Apollo 2*, and the name *Apollo 3* was given retroactively to a previous unmanned mission. The next three Apollo flights were unmanned test flights. *Apollo 7* was the first Apollo flight with astronauts on board, and it was launched successfully on October 11, 1968. The three-man crew spent nearly eleven days in orbit and tested the Apollo Command and Service Module (CSM), a large capsule that had been built with enough room to carry three astronauts to the moon. The ultimate plan was for the command module to function as a kind of mother ship, essentially parking in orbit with one astronaut at the controls, while the two other astronauts boarded a smaller spacecraft inside the mother ship that would separate and carry them to the moon. Once there, the two astronauts would exit the smaller spacecraft, called the Lunar Lander, and explore the moon's surface. They then would get back onboard the Lunar Lander, blast off from the moon, reconnect with the mother ship, and all three astronauts would travel together back to Earth. Previous tests had proven the viability of this concept, which required a series of precisely timed steps. And each step had to be tested before the trip to the moon. *Apollo 7*'s mission was to test the command module, and it worked beautifully. For the first time, the astronauts also broadcast live from space, an important feature that would continue to connect a curious public to NASA's space voyages.

Finally, a tumultuous 1968 closed with the December 21 launch of *Apollo 8,* which made the three astronauts on board the first crew to reach the moon. They made it there on Christmas Eve and spent the day in orbit. That evening, they broadcast live, showing stunning pictures of the moon and the Earth. One of the photos (called *Earthrise*), taken by astronaut William Anders as the Earth rose past the moon's surface, would become iconic. When it was time to end the broadcast, they took turns reading the first ten verses of the Book of Genesis in the Holy Bible.

"For all the people on Earth, the crew of *Apollo 8* has a message we would like to send you," Anders said. "'In the beginning God created the heaven and the earth.

"And the earth was without form, and void; and darkness was upon the face of the deep.

"And the Spirit of God moved upon the face of the waters. And God said, Let there be light: and there was light. And God saw the light, that it was good: and God divided the light from the darkness."

Jim Lovell followed with four more verses. Frank Borman read the last two and added, "And from the crew of *Apollo 8*, we close with good night, good luck, a Merry Christmas, and God bless all of you—all of you on the good Earth."

The six-day mission ended successfully with a return to Earth on December 27 and this pronouncement from Lovell: "Roger, please be informed there is a Santa Claus."

During this time, Al Hamer and I were putting in long hours. We'd work our regular shift from eight to five, go home for dinner, and then return to the office at night to check with the engineers from the NASA installation that was monitoring the flight. No one asked us to do it. We had computed the orbits, and we just wanted to be on hand to see that things were going as they should. We did this so frequently that our supervisor finally put a stop to it. We couldn't continue working sixteen hours a day at Langley without permission and without getting paid, he said. We weren't asking for pay. To us, this was fun. But we adjusted our schedules to make sure we weren't working far beyond our stipulated hours.

After two more missions to test the lunar module and its ability to rendezvous with the command module, it was time to walk on the moon. My part was done. I had given it my best, and I was excited about the planned launch on July 16, 1969. But two days before, the television news began showing that Rev. Ralph Abernathy, a confidant of and close adviser to the late Dr. King, had led hundreds of protesters to the front gates of the Kennedy Space Center for a candlelight vigil to protest the launch. Rev. Abernathy, who had succeeded Dr. King as president of the Southern Christian Leadership Conference, had taken on the leadership of King's Poor People's Campaign. The campaign's goal was to draw attention to the poverty and suffering in America's cities, as NASA spent more than twenty billion dollars to send men to the moon. Rev. Abernathy and the protesters returned the next day for a march, led by two mules and a wooden wagon, illustrating the starkness of poverty versus the technological advancements about to unfold just beyond the space center's gates. Rev. Abernathy had requested a meeting with NASA administrator Thomas Paine, who came out to greet the demonstrators. News photographers snapped pictures and took video as the protesters slowly approached, singing "We Shall Overcome." Rev. Abernathy spoke, telling Mr. Paine and the cameras that a fifth of the people in this country didn't have adequate food, clothes, housing, and medical care and that there was an "inexcusable gulf between America's technological abilities and our social injustice." He said the nation's priorities were misplaced and urged that funds used for space exploration be spent to help those citizens suffering from the effects of poverty.

His points were powerful, and Mr. Paine seemed from the news coverage to handle the conundrum well. He showed empathy and humility, acknowledging "the tremendously difficult human problems." He added that "if we could solve the problems of poverty by not pushing the button to launch men to the moon tomorrow, then we would not push that button." He suggested that there might be a way for NASA to contribute to addressing the issues raised, and he asked Rev. Abernathy to pray for the astronauts' safety. Paine even invited some of the demonstrators to watch the launch from the VIP section at the space center. The two men shook hands and parted ways.

The next morning, July 16, 1969, astronauts Neil Armstrong, Michael Collins, and Edwin "Buzz" Aldrin boarded the spacecraft for their journey to make history. At 9:32 a.m., the huge Saturn V rocket—as tall as a 36-story building and at 6.2 million pounds, the weight of about 400 elephants—lifted off, headed to the moon. Thousands of spectators lined the beaches and highways in Florida near the launch site. As I watched the launch with my Langley colleagues that Wednesday morning, I tingled with excitement, confident that my numbers were right. But I prayed silently that there would be no technical glitches. In this line of business, that possibility always hovered.

By the following Sunday, as the lunar lander, dubbed Eagle, made its way to the moon, I was attending a leadership conference for Alpha Kappa Alpha Sorority, Inc., in the Pocono Mountains in Pennsylvania. My sorority had been a constant in my life since 1934, when I followed my friend Dit into the sisterhood. Our favorite high school math teacher, Angie King, who later also taught me math in college, most likely influenced many of the girls at our high school and college to become AKAs. She was supersmart, ladylike, and nurturing. She was the faculty sponsor of the student chapter of the sorority at West Virginia State and a charter member of the Alpha Omicron Omega graduate chapter in the Charleston-Institute area of West Virginia in 1929. I admired everything about her, and I figured if Alpha Kappa Alpha helped make ladies like her, I wanted to be one. So I couldn't wait to get to college to join the sorority.

Later I would come to appreciate the sorority's focus on service. Our motto, "Service to All Mankind," speaks to that mission, and I have enjoyed being part of uplifting our community through scholarship programs, mentoring, voter registration, and so much more for more than eighty years. I especially love the camaraderie and the strong network of college-educated, highly motivated black women. When I first moved to Newport News, I walked into a ready-made community of sorority sisters who helped me adapt and feel at home, and that scenario is repeated every day around the world. I was more likely to see my NASA colleagues Dorothy Vaughan, Mary Jackson, and Eunice Smith at a sorority meeting or a sorority function than at work once I was moved out of the West Computers unit. When NASA dissolved the segregated unit, the remaining women were spread out over various divisions throughout the Langley complex. Since I didn't eat in the cafeteria at work, I rarely had a chance to interact with them or anyone outside of my division at work.

My oldest daughter, Joylette, joined the sorority at Hampton Institute in 1960, and one of her "line sisters" (the young ladies who go through a "pledge" process and enter the sorority together) was future NASA superstar Christine Darden. The girls came to think of me as their "room mother," and I even hosted a barbeque for them in the backyard of our home. Little did I know that Christine would come to work at Langley years later in 1967 as a data analyst during NASA's increased recruiting efforts to hire more black employees. When she later showed up with her husband and two daughters at my church, Carver Presbyterian, I didn't recognize her right away. As head of the choir, I walked over and introduced myself to her, welcomed her to the church, and invited her to join the choir. That's when she reminded me of our connection during her college days. Christine became a faithful member of the choir, and we spent good time together outside of work through the choir and our sorority.

It felt completely natural that I would be surrounded by my sorority sisters on July 20, 1969, for what could become one of the most important moments of my career. And no matter how cool I may have looked on the outside, I could not yet relax. Things could go spectacularly well, which I fully expected, but I knew, too, there was a good chance of catastrophe. So between meetings I found a television to keep up with the crew's progress. As the time for the landing drew near, I sat, surrounded by my sorority sisters, in front of the small black-and-white television for this historic moment. The presence of my sisters was comforting.

At 4:17 p.m. EDT, a voice cracked from the lunar module. It was Armstrong, reporting, "The Eagle has landed." My sorority sisters celebrated with applause, but I knew better than everyone around me that this was just the first of many things that had to go right for the mission to be a total success.

Hours more would pass, and I was back in my room at the Hill-side Inn when the astronauts finally emerged from the spacecraft. I was among the estimated six hundred million people worldwide watching via television at 10:56 p.m. as Armstrong stepped off the ladder and became the first man to walk on the moon.

All of us who were listening and watching then heard him make an iconic proclamation: "That's one small step for man, one giant leap for mankind." (Armstrong later would reveal that he actually said "one small step for *a* man . . ." but that the word was inaudible.)

Aldrin followed him out onto the moon's surface nineteen minutes later, and they planted an American flag. The flag seemed to wave perfectly (though we'd learn later that engineers had designed a special, retractable flagpole, since there is no breeze on the moon). Pride filled me—pride in my country, pride in the astronauts, pride in all the men and women I knew who had worked hard behind the scenes to give our nation this moment. We had met President Kennedy's call to make it to the moon by the end of the decade.

As exuberant as we all were in that moment, we were not across the finish line yet. Our astronauts had to get back home safely, and that required yet even more precision and timing. This was not like missing your airline flight and getting stranded at the airport for, say, six hours, until you can catch the next plane. Our astronauts had just a small window of time to connect with that orbiting vehicle going around. We had done the error ellipsoid calculations, and we knew that if they missed it by a certain number of degrees or more than certain feet per second, they were done. There would be no way for them to get back.

Fortunately, everything else worked as planned. Armstrong and Aldrin spent about two and a half hours performing various tasks, which included unveiling and leaving on the moon's surface a plaque that read: "Here men from the planet Earth first set foot upon the Moon July 1969, A.D. We came in peace for all mankind." The stainless-steel plaque had the engraved signatures of President Richard Nixon and the three astronauts. They also took photographs, collected 48.5 pounds of lunar rock and soil samples, and deployed four experiments. Other items that the astronauts left behind on the moon included commemorative medallions bearing the names of the three deceased *Apollo 1* astronauts and two cosmonauts (Russian space travelers) who also had died in accidents, as well as a tiny silicon disk containing microminiaturized goodwill messages from seventy-three countries and the names of congressional and NASA leaders. The astronauts then returned to the lunar module and closed the hatch for about seven hours of rest. In all, Armstrong and Aldrin spent about twenty-one hours and thirty-six minutes on the moon before blasting off and docking again with the command service module for their journey back to Earth. On Thursday, July 24, 1969, at 12:50 p.m. EDT, the command module splashed down in the Pacific Ocean about 920 miles southwest of Honolulu with all three astronauts safely inside.

Those bright, colorful parachutes were quite a beautiful sight.

We had done it. We had come from behind for the strongest possible finish. We not only beat the Soviets to the moon, but also all three of our men made it back to Earth alive. The Space Race was done, and our men had planted the victory flag. I was part of the team that helped them win. This was more than I ever could have imagined as that little girl who once loved counting the stars. And the long-ago fascination with the wild idea of becoming a research mathematician had led me higher than I even knew was possible. By the grace of God, though, I'd figured it out, just as Dr. Claytor had challenged. The magnitude of the moment washed over me.

And finally, I could take that deep sigh of relief.

chapter 11

LAND ON A FIRM FOUNDATION

My career at NASA continued long after *Apollo 11*'s first thrilling mission to the moon. I would assist with five more Apollo flights that also made it to the moon and gave our astronauts a chance to expand our nation's knowledge of the world beyond. I was happy to be part of it all.

By the mid- to late 1960s, though, the space program was competing for federal dollars with the Vietnam War, President Johnson's War on Poverty, and other big-ticket programs. Once the Space Race was won, public interest in NASA's moon voyages also seemed to wane. *Apollo 17*'s twelve-day mission in December 1972 brought that historic program to an end.

Before then, my buddy Al Hamer and I had hoped that NASA would build on its successes to the moon and focus next on exploring Mars. The two of us even wrote a research paper, doing some preliminary work for an Earth-to-Mars trajectory. The paper, titled "Simplified Interplanetary Guidance Procedures Using Onboard Optical Measurements," was published in May 1972. A human mission to Mars had been included in a bold plan developed by the Space Task Group appointed by President Nixon in 1969 to chart a post-Apollo course for NASA. The task force also had envisioned a low-orbiting space station with a permanent crew and a space vehicle that would transport passengers and materials easily from Earth to support it. But the most ambitious parts of the plan would not prevail right away. President Nixon announced plans on January 5, 1972, to move forward with the more cost-effective space shuttle:

> I have decided today that the United States should proceed at once with the development of an entirely new type of space transportation system designed to help transform the space frontier of the 1970s into familiar territory, easily accessible for human endeavor in the 1980s and '90s. This system will center on a space vehicle that can shuttle repeatedly from Earth to orbit and back. It will revolutionize transportation into near space, by routinizing it. It will take the astronomical costs out of astronautics. In short, it will go a long way toward delivering the rich benefits of practical space utilization and the valuable spinoffs from space efforts into the daily lives of Americans and all people.

My work shifted to calculations for the space shuttle missions. Though it seemed our most exciting days may have been behind us, I still felt fortunate to have a job. The early 1970s were a time of significant reorganization, cost-cutting, and layoffs. As far as I know, my name never made it onto any of the reduction in force (RIF) or reduction in grade (RIG) lists. But the reorganization had an indirect impact on one of the colleagues I most highly respected: Dorothy Vaughan, the former head of the West Computers, who had transformed her career by becoming a computer programmer. She was on the list to be promoted, not laid off. But the job was not enough to keep her at Langley any longer. In 1971 she retired, ending a career that had set so many black women, including me, on a path to success.

My young friend Christine Darden made it through the first round of cuts in 1970, but two years later she learned that she was on the list for another planned round of layoffs. Seniority ruled, and Christine noticed she had been bumped onto the list by a black male employee who had been hired with her at Langley. But he had been assigned to an engineering group and promoted at least a couple of times, while her career had remained stagnant. Christine mustered the courage to go to her division chief with a pointed question: why were the male employees who came to Langley with the kind of credentials she had (a master's degree in math) placed in engineering groups, while the women were assigned to the computer pools with no plans for advancement? The chief responded that none of the women had ever complained before. Christine not only saved her job, but also, two weeks later she got a new assignment that would change the trajectory of her career. She was placed in a group that was researching the sonic boom, an explosive sound produced when an aircraft moves faster than the speed of sound. Immediately her new supervisor gave her a big assignment that took three years of research and led to the 1975 publication of groundbreaking research that still is used in the industry. Christine returned to school while working full-time and raising her daughters, and ultimately earned a doctorate in mechanical engineering. With the help of a strong advocate in NASA's Equal Employment Opportunity Office, Christine was promoted to the agency's Senior Executive Service, the highest level for a federal employee. And she has become the preeminent international expert on the subject of sonic boom. I beam like a proud parent when I talk to schoolchildren about her intelligence, tenacity, and fearlessness. I especially love that Christine remains as humble and down-to-earth as when I first invited her to join our church choir.

The years at NASA passed quickly. By the 1980s I had learned Fortran, the language used to program the early generation of computers, and I was using the computer regularly in my work. After all this time, I still loved my job. More of my peers, though, had begun winding up their careers at Langley. When my close friend Eunice Smith decided in 1984 that she was ready to move to the next chapter in her life, I began contemplating it, too. Jim and I wanted to travel, spend more time with our families, and have more control over our time. The next year, Mary Jackson, my former West Computer colleague who broke barriers in 1958 by becoming NASA's first black engineer, also decided to retire. After rising as high as she figured she would be allowed to go in 1979, she had made the tough decision to become the federal women's program manager in the Human Resources Department, where she helped to steer the careers of talented young women of the future. Mary spent her last years at Langley as an equal opportunity specialist, pushing others through the glass ceiling that had trapped her.

My longtime coworker and friend Al Hamer and I were planning to retire in 1985 as well, and we had even started working on our retirement papers. But one day our division chief stopped by and asked our friend engineer John Young to chair a committee that was studying the use of solar panels in the planned space station. The space shuttle had been completed, and President Ronald Reagan announced plans in 1984 finally to build the space station as the next logical progression in space. John told him that he didn't want to chair the committee, but the chief assigned him to serve as a member. John then asked the chief if I could be assigned to work with him. The chief then turned to me and asked, "Do you want to work for John?"

"Sure," I responded. "Any day of the week."

I had been there from the agency's first venture into space, and staying to work on the space station seemed a good place to end my career. I put my retirement papers aside for the moment, and so did Al. The three of us—John, Al, and I—would work together at Langley for fifteen years, and we had quite a wonderful time. Our lunchtime bridge games were so competitive that we sometimes drew onlookers. Al even trusted me enough to let me cut his hair. In the 1970s, many men began wearing longer hair styles. But when Al's hair got a bit too long for a professional, polished look, I offered one day to give him a good trim. He agreed, brought a pair of barber scissors to work one day, and we ripped off a large piece of paper from the ream that fed through the old computer and spread it on the floor around his desk. I cut his hair right there. Al was happy with my handiwork, so I became his unofficial barber. And he wasn't my only "client" in the office. A contractor named Daniel Giesy, who had been added to our unit in 1977 to assist me in running a computer program, showed up at the office one day with his hair looking a bit too scraggly. I offered to give him a trim as well. Daniel, then the new guy in our office, probably was a bit surprised by my offer, but he agreed, was satisfied, and became another regular. I was happy to assist those guys in maintaining a professional look, and they got free haircuts out of the deal.

I genuinely liked my coworkers, and I cannot recall a single day when I did not want to go to work. But January 28, 1986, was one of our toughest days at the office. Like the rest of the world, my colleagues and I gathered around a television screen to watch the launch of the *Challenger* space shuttle. But our excitement quickly turned to horror when just seventy-three seconds into the launch a thick plume of white smoke and then an orange flame indicated that something terrible had gone wrong. We gasped and stood silently in shock when we learned shortly afterward that the space shuttle had exploded and killed all seven crew members. The news was even more heartbreaking because we knew children around the world had tuned in to the live broadcast to see Christa McAuliffe, a social studies teacher, travel into space for the first time. She had been selected from thousands of applicants for a special NASA program, the "Teacher in Space Project," aimed at inspiring students to explore the science, math, and technology fields that had made the voyage possible. Christa was scheduled to teach lessons live from space to millions of schoolchildren and even give them a tour of the space shuttle. At NASA we knew the risks every time an astronaut climbed aboard a spacecraft. But the children didn't know. The general public didn't know. What would this catastrophic end say to them about our space program? It was a quiet, somber day at the office.

I thought about Christa's nine-year-old son and six-year-old daughter, who were watching at the Cape Canaveral launch site. I thought about all of the other crew members who had died and their family members, many of whom also were spectators at the Florida launch. My own heart had filled with such pride as I watched Ronald McNair, one of the first black astronauts, and Judith Resnik, one of the first women on the job, climb aboard the spacecraft. Astronaut Sally Ride had become the first woman to journey into space on a previous June 1983 Challenger mission, and astronaut Guion Bluford followed her two months later and became the first black astronaut on another Challenger mission. But all four trailblazers had been selected together in Astronaut Group 8, which included thirty-five space travelers and was at the time NASA's largest and most diverse group. In addition to McAuliffe, McNair, and Resnik, the others who died in the *Challenger* disaster were astronauts Francis R. Scobee, Michael J. Smith, and Ellison Onizuka, as well as Hughes Aircraft engineer Gregory Jarvis. I will never forget the men and women who died on that sad day.

The year 1986 would end on a happier note for me, though, as I followed up on my plans to retire from NASA during the summer after thirty-three years. I describe it as a happy time because I had accomplished more than I ever imagined was possible, and I was leaving in good health with the freedom to do whatever I desired. NASA held a retirement luncheon and presented me some awards, which I cherish, but I wouldn't let the girls give me a big retirement celebration. I didn't even tell Kathy and Joylette about the luncheon because I really didn't want to make a big fuss. Jim was there, of course, and I also invited Connie, who by then had moved back to Hampton from New Jersey with her children. Once I had decided to retire, I moved on in my mind to what was next, so I didn't need to spend any time looking back. I may not have known exactly what lay ahead, but I knew that my foundation—my faith, my family, and my friends—would remain solid the rest of the way. And that was enough.

As I was retiring, my oldest granddaughter, Laurie, was graduating from Hampton University with a degree in business, and her younger brother, Troy, was a student there. Our family is full of Hampton graduates, so the university's annual Homecoming remains one of my family's biggest celebrations of the year. All six of my grandchildren grew up attending Hampton University's Homecoming festivities year after year because their parents, all Hampton graduates, rarely missed the big weekend. All Historically Black Colleges and Universities (HBCUs) across the country have some version of this reunion weekend. When their graduates leave their alma maters and go out into the wider world, they appreciate more than ever what they got there. They not only got a good, cost-effective education, but also they became adults in a place that made them feel seen, heard, and valued. The best of them were given opportunities and the confidence to lead. It's no wonder that many of this country's greatest black leaders are products of HBCUs.

My youngest grandson, Michael, is a perfect example. He is an intelligent young man who grew up in Orange, New Jersey, and graduated in 1987 as number three in his high school class. He loves math and science, like me, and wanted to be an engineer. He accepted a scholarship to an overwhelmingly white institution, but he was one of just three black undergraduates in the entire engineering school. By his sophomore year, he was the only black student in the program. He felt isolated in an environment that did not feel welcoming, and he lost his confidence. His grades dropped, and he ultimately left the university. I had a conversation with him to encourage him not to give up. He had too much potential, and I let him know that he had other options. I didn't push him to transfer to Hampton, but it seemed a natural transition for him to make. He likes to say that Hampton University was his place of restoration. He thrived there, graduating in 1995 with a degree in computer science. Now a married father of four, Michael is close to completing law school.

Hampton is at the center of so much of my family's success. Laurie, independent like me, followed her own path and became an entrepreneur who established her own nail salon business that has flourished for more than thirty years. Troy, who graduated from Hampton with a degree in finance and math in 1988, worked as a teacher and math coach for fifteen years and as a night auditor at a hotel. This is why we celebrate Hampton University.

No matter where my girls were living throughout their adult

lives, they usually made their way home with their families to Hampton for Homecoming. We cheered for the Pirates at the football game, attended the parade, scholarship dinners, and sorority and fraternity step shows on campus, and drove to get-togethers at the homes of alumni throughout the Hampton Roads area. When my grandchildren were kids and Homecoming weekend fell on Halloween, they would spend the day piecing together their costumes and then launch from my house to go trick-or-treating in the evenings without their parents. Though I am no longer able to participate, Homecoming is still a time for my family to connect with old friends, celebrate the place that has given us so much, and just have a good time.

I was happiest at our big, family get-togethers. Jim and I often hosted the family's Fourth of July cookout. Our backyard on Mimosa Crescent would be packed with family members and friends gathered around card tables covered by newspaper and piled high with boiled crabs and corn on the cob. I wore my apron and made a steady trek to and from the kitchen to refill the bowl of potato salad, trays of deviled eggs, and pans of fried chicken. I'd pause when someone pulled out a deck of cards for a game of pinochle. My grandchildren ran around the yard, tagging one another, playing. And suddenly, it seemed, they were teenagers, and then college students and adults, gathered around their own tables. I relished all of their successes. My grandson Gregory served proudly in the US Air Force for twenty-five years; his brother, Douglas, attended culinary arts school in Baltimore and became a popular chef; and their sister, Michele, became a middle school math and science teacher in New Orleans.

When it was another family member's time to host a family gathering, the scene looked the same, just in a different place. Those moments added consistency to the years. The old folks just kept getting older, and the kids grew up and produced another generation of children running around. But there was always our Homecoming, Fourth of July, Thanksgiving, Christmas, and family members arguing over who made the best deviled eggs or potato salad.

I've never been the touchy, feely type, but my grandchildren always knew they could call me to help with a complicated word problem. Sometimes I'd even call and challenge them with one. When Laurie first learned how to add three-digit numbers, she loved to phone me and race me to finish. She would give me the numbers, and I'd add or subtract them quickly in my head, while she scrambled on paper to beat me to the answer. I didn't believe in letting kids win, but she had fun trying.

I kept on the kitchen table in my house a deck of cards, crossword puzzles, and games that would keep all of our minds sharp. And whenever my children and grandchildren came to my house, they knew they would find a hot pot or pan on the stove, filled with homemade soup, or a casserole, or fried chicken and the fixings, rolls or cornbread, and of course their favorite homemade applesauce. I always used fresh Granny Smith apples and enough sugar to give my applesauce the perfect blend of sweet and tart. Friends and family members requested this specialty so often that I sometimes put it in jars and gave them away as Christmas gifts.

My apron was probably the most well-worn garment in my house. I loved cooking for my family and friends and watching them enjoy my food. But I can be a real stickler, too, when it comes to manners. My four grandsons can attest that one of my biggest pet peeves is a man wearing a hat indoors, especially at the dinner table. I taught my grandsons that a gentleman always removes his hat before entering a building. And no man, young or old, would be allowed to enter the door of my home without that respectful gesture. On the rare occasion when one of my grandsons momentarily forgot and came dashing inside with his head covered, I'd raise my eyebrow and give him "the look," my silent way of asking, "Have you lost your mind?" And it was usually enough to get the desired compliance right away.

Retirement gave me the opportunity to create more of those memories and to spend more time speaking to schoolchildren. So many students are intimidated by math and science, but I reminded them that math and science are everywhere, from the most basic recipes of their favorite foods to designing a video game or a spaceship. My grandson Troy, the math coach, came up with an idea to invite me to speak at his school in Camden, New Jersey. This was in 2003, long before the *Hidden Figures* book and movie. He was concerned that his students seemed obsessed with fame and careers such as sports and modeling. He played into their fascination with fame and told them to look me up on Google (imagine that!). By then, a few stories had been written about my contributions to NASA, and the students got all excited, asking him, "Mr. Hylick, is that really your grandmother?" They had no idea how excited I was to meet *them*. I talked to them about the importance of getting an education and how education and a lifelong fascination with learning new things had helped me to achieve my dream of becoming a research mathematician. I told them that we use math in everyday life and that it's easy. I'm not sure they believed that part, but they seemed engaged. And just maybe I sparked an interest in at least one of them. That's really all it takes sometimes to ignite a dream.

One of the first times that I went into the classroom to talk about my career was in the early 1970s. I was still working at Langley, and one of my former teachers, Mrs. Leftwich, invited me back to my hometown, White Sulphur Springs, West Virginia, to talk to students. At the time, NASA had recently launched its first space station, and to make my presentation fun for the students, I did some calculations before I got there and pinpointed the precise moment when everyone would be able to see the space station in White Sulphur Springs. Many people didn't even know there was an actual space station up there. They may have thought it was a flying star or just didn't see it. But I told them what time they could see it and how far up it was, and they were in awe. Later that evening I went outside and looked up at the time I'd told the students to look for it, and there it was, glinting in the sky.

Decades later, I was speaking at a college in North Carolina, and it occurred to me that it would be really interesting for the students to hear from an actual astronaut. I had appeared on a couple of programs with a gifted young astronaut named Leland Melvin and we'd become friends, so I called him. There was no answer, but I was just about to go onstage in the first classroom when my phone rang. It was Leland, and I told him what I wanted. "Well, put them on," he said. We put him on speaker, used a microphone, and he talked to the students. They were so excited that they got to talk to a real astronaut. I think they were really impressed that I could call an astronaut and he would answer. I hit the jackpot on that one.

That has been one of the most beautiful things about my post-NASA years, seeing the increasing diversity among the younger astronauts. I met Dr. Mae Jemison, who became the first black woman to travel into space, in 1992, when she spoke at an Alpha Kappa Alpha event many years later. I was delighted when President Obama named Major General Charles F. Bolden Jr. as NASA administrator in 2009. And I've even gotten to know a couple of the black astronauts personally. Leland and I appeared on several programs together and talk regularly. He is an extraordinary young man and terrific role model, particularly for youths focused on sports. Leland was a star wide receiver at the University of Richmond and was drafted into the NFL by the Detroit Lions in 1986. A hamstring injury sidelined his professional football career before he got to play in the league, but with a bachelor's degree in chemistry, he instead pursued a career in science. He got a job in the Nondestructive Evaluation Sciences Branch at NASA Langley in 1989, earned a master's degree in materials science engineering from the University of Virginia in 1991, rose through the ranks at Langley, applied to become an astronaut, and was selected in 1998. He flew on space shuttle flights in 2008 and 2009 to the international space station. My conversations with him have been fascinating.

During one of our talks, we discussed a research paper that Al Hamer and I had done in the 1960s about how to use the pattern of the stars as a guide to navigate from space back to Earth if a spacecraft lost power. Some media reports have incorrectly credited that study with helping the troubled *Apollo 13* flight to the moon find its way back home after an explosion and power outage in space. But quick thinking, ingenuity, and another scientific method actually helped the astronauts survive that cold, perilous journey back to Earth—a journey that was dramatized in an Academy Award–winning movie starring Tom Hanks. In our research paper, Al and I had laid out the patterns of the constellations, which we believed could be used as navigation points if ever needed. But Leland explained that the view of the stars from Earth is different from the eye level view in space. The stars are indistinguishable when you're flying among them, he told me. Thus the information in the study was not useful to them. Leland's perspective was so informative. It would have been great to have had that kind of connection to the astronauts while I was still researching and writing.

Astronaut Yvonne Cagle also has become like family.

She and Joylette connected first in 2016, when Joylette and her son, Troy, attended an event sponsored by the black avaiators' group, Shades of Blue, to receive an award on my behalf. Yvonne gave her an autographed photo to give to me, and she called me a short time after that. We began talking regularly by phone, and she said she felt horrible that she had allowed so much time to pass and that she had missed out on my mentorship. We immediately began trying to make up for lost time with what became regular Sunday afternoon chats.

The first time the two of us saw each other face-to-face was quite special. Rehema Ellis, then a television reporter based in New York, had come to interview me in Newport News, and I didn't know Yvonne was going to be there. She slipped in quietly, and Rehema motioned for her to come in and say hello. Yvonne was wearing her blue astronaut suit, and when I saw her, it took my breath away. There, standing in front of me, was a real, live Lieutenant Uhura. But this was not *Star Trek*. This was real life, my life. She reached for my hand. I reached back, and we just looked at one another silently. In that moment, the past, the present, and the future came together. My daughters and the journalists felt it, too. There were no dry eyes in the room.

By the time I met Yvonne that day, life had begun to change dramatically for me. The years have a way of doing that. They speed up on you, slow you down, and sometimes knock you to the ground. One day you're eighty years old, driving across the country on a girlfriends' trip to Ohio, as my friend Mae Barbee Pleasant and I did in about 1998. Then you're eighty-eight years old, taking boxing lessons to stay in shape, as I did in 2006. And then you're blindsided and have to figure out how to drag yourself up and go on.

In 2010 I lost my middle daughter, Connie. She was my alter ego, the free spirit of the bunch, the one who was untethered to places and things, who would follow her heart wherever it led. She always loved children, and when she was in college, we joked about it, calling her the "Pied Piper" because she babysat the professors' kids and always had a string of children following her. That love for children and connection with them made her an excellent kindergarten teacher. But when the system dropped its heavy boot on her neck, began choking out her creativity, and forced her to shift to upper grades and teach toward standardized tests, she tapped out. She quit her steady teaching job and persuaded Jim and me to support her new business venture, Connie's Trucking, by purchasing an eighteen-wheeler. She learned how to drive the big rig and spent years exploring the country in it, sometimes in the driver's seat.

Connie drifted back into teaching and made New Orleans her home for a while in the 1990s, when she went to help care for her friend, Elsie, who was ill. A friend of Elsie's was a school board member, who was instantly charmed by Connie, and offered to get her a teaching job. Out of the blue one day, I received a call from my daughter, saying, "Mom, I think I'm going to stay in New Orleans." That was Connie.

Eventually, Connie returned to her home in Hampton to be closer to Jim and me, and she faithfully attended to us, coming over to the house to cook, plant flowers, decorate the Christmas tree, whatever was needed. When Jim reached his mid- to late eighties, Connie insisted on becoming our designated driver. "Just call me whenever you need to go somewhere," she said. Then, quite suddenly in April 2010, she started feeling badly and had to be hospitalized. Having hidden most of her discomfort from us, she didn't get a good report from the doctors. They let us know her condition would not improve. She was released on April 17, and her good friend Gail converted the great room in her home into a room for Connie. The two had been housemates previously and were as close as sisters. My family will forever appreciate her dedication to Connie.

Cards and flowers filled Connie's room, our sorority sisters brought food, and there was a constant flow of relatives and friends to visit her, including the Core twins from White Sulphur Springs, Annette and Janette, and cousins from St. Louis, Houston, and New York. Kathy and Joylette took turns staying with Connie at night, while Gail and Elsie, whom everyone called "Bunny," took care of her during the day. Connie's children—Gregory, Douglas, and Michele—also stayed close. Connie passed away on May 4, 2010, one of our saddest days.

I'm still mad at Connie for leaving us so soon. But I find comfort knowing that she lived each day like it would be her last. By then I had experienced many other painful losses, including Mamà, who died at age eighty-three in 1971, and Daddy, at ninety-one in 1973. Sister died in 2002, about a year after I persuaded her to move from White Sulphur Springs to Hampton, where we had lots of family. By then she had earned her master's degree and had retired as a teacher after forty years. Many of my friends had died too, including my best friend from college, Constance (Dit). When we lost our daughter, I was was ninety-two and Jim was eighty-four. And sure enough, life seemed to slow us down.

In 2012 we made a big move into an assisted living retirement community. Jim and I hated to leave our home on Mimosa Crescent. How would we ever squeeze our life's accumulation of stuff into a two-bedroom apartment? How would we adjust to losing so much of our independence? Jim was a proud man. I'd watched him jog around Hampton University's campus and in the Homecoming parade with those young cadets until he was well into his seventies. The university's ROTC program finally nudged him into retirement with a huge celebration when he turned eighty, perhaps because they knew he would never leave on his own. Jim also worked as a substitute teacher in Newport News until his eighties. Many mornings he'd get up at four o'clock to get dressed and wait for a call from one of the schools that needed a substitute teacher, and he would stay until the last person left. The kids kept him young and made him feel needed and appreciated. Eventually, though, my former boxing trainer Vernon Lee became Jim's driver, taking him to the schools and to choir rehearsal. I marveled when Jim dropped to the floor to do sit-ups and push-ups to keep his washboard abs. The body held up, but the mind started to slip, a little forgetfulness here and there at first. And the girls grew worried.

I've always been pretty pragmatic, so we adjusted to our new home. We've danced on with the help of our wonderful daughters, who have made us so proud. Kathy graduated from Hampton, went on to get her master's degree, became a teacher for many years and then a high school counselor. Joylette returned to work after her children started school and held jobs in computer programming, system integration, and project management for several companies, most recently Lockheed Martin. Both are now retired, and they travel frequently from their homes—Kathy in North Carolina and Joylette in New Jersey—to visit. We have devoted caretakers and a steady stream of friends. My sorority sisters come every week to play bridge or bingo with me (and they think I don't know when they let me win). We sing our sorority hymns, and I raise my hands and direct them, like old times. There's a favorite part in one of the songs that says, "When cares of life o'er take us and our locks are turning gray, we'll always reverence A.K.A. forever and a day." I reach up with a smile and pat my own white hair.

By the time the recognition came from President Obama and the White House in 2015 and later the Oscars and building dedications, Jim was unable for health reasons to make the trips with us. But when we got home the girls and I couldn't wait to tell him all about what happened. He and I were both amazed by all the hoopla after the book and movie the following years. When I'd come back home from one event or another, Jim would chuckle and look at me as if to say, "Can you believe this, Kid?"

Kid. That was his nickname for me when it was time for us to retire to our rooms. Our aides lined our wheelchairs facing one another for our nightly fist bump. No, I could hardly believe this remarkable journey, I thought, as our fists touched night after night.

"Good night, Kid," he said.

And I smiled.

ACKNOWLEDGMENTS

We are grateful that our mother lived for 101 years, but she did not make it to see the publication of this memoir. She died on February 24, 2020, while it was being written. We would have to write another book to thank all of the people who brought comfort and joy to her and her devoted husband, Jim, during their later years. He preceded her in death on March 13, 2019. Thank you all for honoring and recognizing Mom and her work. She was truly humbled, grateful, and even a bit bewildered to the end by the tremendous outpouring of love, kindness, and respect.

We are particularly thankful to the entire team at HarperCollins Publishers for seeing this book to fruition, especially our editor, Tracy Sherrod, editorial director at Amistad. This project would not have been possible without our formidable team, which includes our legal representative and friend, Donyale Reavis, who is tenacious in her protection of us; our literary agent, Jennifer Lyons; and writer Lisa Frazier Page.

Thanks to author Margot Lee Shetterly, whose book *Hidden Figures* helped bring to light the amazing story of the African

American women who worked as human computers and whose many detailed interviews of Mom preserved precious memories. We applaud the administrators and staff at NASA for the many beautiful ways you have acknowledged our mother in recent years, particularly astronauts Dr. Yvonne Cagle and Leland Melvin, who both befriended Mom and accompanied her to many events, as well as our dear friends, Dr. Christine Darden and Michael Chapman (former president of the National Technical Association). Many others also shared their memories to add flesh to some of the stories that Mom recalled for the book, including our children Laurie Hylick Braxton, Troy Hylick, and Michael Moore; Aunt Patricia Goble Evans; our former West Virginia neighbors Annette Core Henderson and Janette Core Henderson; Mom's dedicated sorority sisters Maggie Macklin and Dianne Blakeney; and her former coworker Daniel Giesy. To every family member and friend, you know how much you meant to Mom. To our niece and nephews (Connie's children)—Michele Boykin Sanders, Greg Boykin, and Doug Boykin—you were part of Mom's heartbeat.

We cannot say enough about Joseph Saunders, chief of police at West Virginia State University, who went beyond the call of duty to provide security for Mom and our family and watched over her with the protective eyes of a son for most of our travels and appearances; nothing we asked of him was too much. Likewise, Dr. Fanchon Glover blessed us with her superior planning skills while orchestrating Mom's perfect 100th birthday celebration in West Virginia. "Chon" became part of our special "K-Team," which also included Chadra Pittman, and Shauna Epps, who planned the final celebrations of Mom's life. And the eulogy delivered by Dr. Brian Blount, president of Union Presbyterian Seminary and Mom's longtime former pastor, was not only heartfelt and personal, it was a literary masterpiece.

The pastors and members of Carver Presbyterian Church for the nearly sixty-eight years she worshipped and served there (from 1952 to 2020) were Mom's second family. The extraordinary efforts of the staff at the Hidenwood Retirement Community helped Mom and Jim live their final years in peace with a good mix of fun, and we always knew they were in great hands with aides Lucy Livingston, Jean Keita, and Derek Robinson.

Finally, we are grateful for the work of organizations, such as the National Visionary Leadership Project and The HistoryMakers, whose extensive digital interviews of Mom and others who left their imprint on the world are absolutely priceless. Those treasured interviews helped make it possible for us to finish this book.

—Joylette Hylick and Katherine Moore

NOTES

introduction: AN UNIMAGINABLE CENTURY

1 *sliced bread, which didn't become one of the century's great inventions until 1928:* Jennifer Latson, "How Sliced Bread Became the 'Greatest Thing'," last modified July 7, 2015, https://time.com/3946461/sliced-bread-history/.

2 *a pandemic that would claim an estimated 50 million lives over the next two years:* "1918 Pandemic (H1N1 Virus)," Centers for Disease Control and Prevention (CDC), last reviewed March 20, 2019, https://www.cdc.gov/flu/pandemic-resources/1918-pandemic-h1n1.html.

2 *a total of 675,000 people died in the United States:* "1918 Pandemic," CDC.

2 *Ford Motor Company also was selling its popular Model T for about $350:* "Ford Model T (1908–1927)", myAutoWorld, https://myautoworld.com/ford/history/ford-t/ford-t.html.

chapter 1: NOBODY ELSE IS BETTER THAN YOU

16 *Ingleside Seminary in Burkeville, Virginia, which was built in 1892 to educate colored girls:* Bernard Fisher, "Ingleside Training Institute," The Historical Marker Database, last revised June 16, 2016, https://www.hmdb.org/m.asp?m=31042.

17 *Covering thousands of acres—eleven thousand at present:* "About the Greenbrier," Greenbrier, May 5, 2020, https://www.greenbrier.com/about-us.aspx.

17 *For most of its existence, the resort was owned by the Chesapeake and Ohio (C&O) Railway Co.:* "About the Greenbrier," Greenbrier.

17 *The property also includes ninety-six guest and estate homes:* Greenbrier, "About the Greenbrier," Greenbrier.

19 *President Woodrow Wilson created the US Food Administration to manage the conservation, distribution, and transportation of food during wartime:* "Years of Compassion 1914–1923," Herbert Hoover Presidential Library and Museum, accessed May 5, 2020, https://hoover.archives.gov/exhibits/years-compassion-1914-1923.

19 *Hoover appealed to Americans . . . wheatless Wednesdays:* "Years of Compassion," National Archives.

20 *There were three registrations for the draft:* "World War I Draft Registration Cards," National Archives, Military Records, NARA microfilm, revised December 2010, https://www.archives.gov/files/research/military/ww1/draft-registration/selective-service-cards.pdf.

23 *The Shawnee tribe originally inhabited the forest near the springs:* John Lund, "White Sulphur Springs, West Virginia," *Geo-Heat Center Bulletin*, May 1, 1996, https://www.researchgate.net/publication/251211950_White_sulphur_springs_West_Virginia.

chapter 2: EDUCATION MATTERS

32 *The devastating 1857 ruling, known as the "Dred Scott decision":* "Dred Scott Case: The Supreme Court Decision," PBS, *Africans in America*, https://www.pbs.org/wgbh/aia/part4/4h2933t.html.

32 *As early as 1837, a benevolent white philanthropist established the Institute for Colored Youth:* "History of Cheyney University," Cheyney University, https://cheyney.edu/for-parents/history-traditions/.

32 *Those federal laws are perhaps better known as the Morrill Acts, named in honor of Justin Morrill . . . both legislations:* Keith Randall, "The Morrill Act Still Has a Huge Impact on the U.S. the World," *Texas A&M Today*, July 2, 2020, https://today.tamu.edu/2020/07/02/the-morrill-act-still-has-a-huge-impact-on-the-u-s-and-the-world/.

33 *The location of the school in the community of Institute, just outside Charleston, was heavily influenced by Educator Booker T. Washington:* "West Virginia State University," Wiki2, https://wiki2.org/en/West_Virginia_State_University.

34 *When he arrived at the Institute, less than half of the twenty-four faculty members had college degrees:* Ancella R. Bickley "John Warren Davis," e-WV: *The West Virginia Encyclopedia*, October 15, 2012, https://www.wvencyclopedia.org/articles/1717.

38 *It dates back to a rich, white slaveowner and the enslaved woman with whom he shared his life:* James A. Haught, "Institute: It Springs from Epic Love Story," West Virginia Department of Culture, Arts, and History, http://www.wvculture.org/history/journal_wvh/wvh32-2a.html.

41 *Miss Turner earned her master's degree in chemistry from Cornell University in 1931:* Sibrina Collins, "Angie Lena Turner King

(1905–2004)," March 13, 2012, https://www.blackpast.org/african-american-history/king-angie-lena-turner-1905-2004/.

43 *In some parts of the state, the unemployment rate was as high as 80 percent:* "The Great Depression," The West Virginia Encyclopedia, e-WV, https://www.wvenclyclopedia.org/articles/2155.

45 *The teenager had been falsely accused a month earlier of raping a white girl in Maury County:* "Oral History Interview with John Hope Franklin, July 27, 1990," Interview A-0339, Southern Oral History Program Collection (#4007) in the Southern Oral History Program Collection, Southern Historical Collection, Wilson Library, University of North Carolina at Chapel Hill, https://docsouth.unc.edu/sohp/A-0339/excerpts/excerpt_786.html#fulltext.

49 *A young W. W. Schieffelin Claytor entered Howard as an undergraduate in September 1925:* Karen Hunger Parshall, "Mathematics and the Politics of Race: The Case of William Claytor (Ph.D., University of Pennsylvania, 1933)," *The American Mathematical Monthly* 123, no. 3 (2016): 214–40, accessed April 5, 2020, https://www.jstor.org/stable/10.4169/amer.math.monthly.123.3.214.

chapter 3: A TIME FOR EVERYTHING

58 *Named in honor of Francis Marion, an American Revolutionary War hero, the town is the county seat of Smyth County:* "Marion Historic District," National Park Service, https://www.nps.gov/nr/travel/vamainstreet/mar.htm.

58 *About five thousand people called Marion home in 1937:* "Marion, Virginia," Historical Population, Wikipedia, accessed April 5, 2020, https://en.wikipedia.org/wiki/Marion,_Virginia.

58 *The park was 1,881 acres of donated land:* Virginia State Parks, stateparks.com, https://www.stateparks.com/hungry_mother_state_park_in_virginia.html.

58 *Another of Marion's local treasures was the three-story, red-brick Lincoln Theatre:* "History of The Lincoln," The Lincoln Theatre, https://www.thelincoln.org/about-us.

60 *It had been named in honor of the Reverend Amos Carnegie, who had come to Marion by 1927 as pastor of Mount Pleasant Methodist Church:* Linda Burchette, "Carnegie Reunion in Marion Brings Students, Teachers Back Together," swvatoday.com, July 19, 2017, https://swvatoday.com/entertainment_life/article_32966602-6c06-11e7-8509-d7f9d4ed5e02.html.

60 *He raised money in the Negro community and secured a grant from the Julius Rosenwald Fund:* "Three State Historical Highway Markers to Be Dedicated in Town of Marion," DHR, Virginia Department of Human Resources, Friday, Nov. 15," Updated December 2, 2019, https://www.dhr.virginia.gov/press_releases/three-state-historical-highway-markers-to-be-dedicated-in-town-of-marion/.

65 *The case stemmed from a lawsuit that had been filed by Lloyd Gaines:* "Lloyd Lionel Gaines," The State Historical Society of Missouri, https://historicmissourians.shsmo.org/historicmissourians/name/g/gaines/.

65 *In 1937 he had talked state officials out of approving legislation that would have granted West Virginia State two million dollars a year over two years:* Albert P. Kalme, "Racial Desegregation and Integration and American Education: The Case History of West Virginia State College, 1891–1973," PhD dissertation, University of Ottawa, 1973.

chapter 4: THE BLESSING OF HELP

72 *Torpedo bombers and other military planes swooped down from the skies:* Patrick J. Kiger, "Pearl Harbor: Photos and Facts from the Infamous WWII Attack," history.com, https://www.history.com/news/pearl-harbor-facts-wwii-attack.

74 *More than a thousand diplomats and family members from Germany, Italy, and Japan were living in and around the nation's capital when the United States entered the war:* Harvey Solomon, "When the Greenbrier and Other Appalachian Resorts Became Prisons for Axis Diplomats," smithsonianmag.com, Feb. 21, 2020, https://www.smithsonianmag.com/travel/when-greenbrier-other-appalachian-resorts-became-prisons-for-axis-diplomats-180974243/.

78 *More than six million Jews were murdered during the Holocaust:* United States Holocaust Memorial Museum, https://www.ushmm.org/learn.

78 *At least a half-million other non-Jews also were executed:* Terese Pencak Schwartz, Jewish Virtual Library, "The Holocaust: Non-Jewish Victims," https://www.jewishvirtuallibrary.org/non-jewish-victims-of-the-holocaust.

78 *Five square miles of the city were leveled instantly, and an estimated eighty thousand people were killed:* "Bombing of Hiroshima and Nagasaki," history.com, updated August 5, 2020, https://www.history.com/topics/world-war-ii/bombing-of-hiroshima-and-nagasaki.

81 *The Norfolk and Western Railway located its headquarters on the West Virginia side in the late 1880s:* "History and Heritage," *Town Square Publications*, http://local.townsquarepublications.com/westvirginia/bluefield/02/topic.html.

85 *Tens of thousands of World War II veterans would develop leukemia*

and other cancers: Atomic Heritage Foundation, Atomic Veterans 1946–1962, June 17, 2019, https://www.atomicheritage.org/history/atomic-veterans-1946-1962.

chapter 5: BE READY

99 *The next year the National Advisory Committee for Aeronautics (NACA, the predecessor to NASA) sent Congress a proposal to double its staff agencywide:* Alex Roland, "The National Advisory Committee for Aeronautics 1915–1958," *Model Research, The NASA History Series*, vol. 1, 1985. https://history.nasa.gov/SP-4103vol1.pdf.

102 *Working in her favor, too, was Executive Order 8802:* Executive Order 8802: Prohibition of Discrimination in the Defense Industry (1941), https://www.ourdocuments.gov/doc.php?flash=false&doc=72.

103 *The first female "computer pool" dates back to 1935:* "When the Computer Wore a Skirt: Langley's Computers, 1935–1970," *NASA History Program Office News & Notes* 29, no. 1 (first quarter 2012), https://www.nasa.gov/feature/when-the-computer-wore-a-skirt-langley-s-computers-1935-1970.

103 *"The engineers admit themselves that the girl computers do the work more rapidly and accurately than they would . . .":* "Hidden Figures and Human Computers," Smithsonian, National Air and Space Museum, https://airandspace.si.edu/stories/editorial/hidden-figures-and-human-computers.

103 *The number of computers grew rapidly:* "When the Computer Wore a Skirt: Langley's Computers, 1935–1970," *NASA History Program Office News & Notes* 29, no. 1 (first quarter 2012), https://www.nasa.gov/feature/when-the-computer-wore-a-skirt-langley-s-computers-1935-1970.

chapter 6: ASK BRAVE QUESTIONS

109 *At the time, just about 34 percent of all women worked outside the home:* "Changes in women's labor force participation in the 20th century," US Bureau of Labor Statistics, *TED: The Economics Daily*, February 16, 2000, https://www.bls.gov/opub/ted/2000/feb/wk3/art03.htm.

114 *the equivalent of about $11,500 in today's dollars:* US Inflation Calculator, https://www.usinflationcalculator.com/.

chapter 7: TOMORROW COMES

126 *a Negro girls' home economics club that would merge years later with the all-white Future Homemakers of America:* "New Homemakers of America, Kentucky," Notable Kentucky African Americans Database, last modified August 8, 2018, https://nkaa.uky.edu/nkaa/items/show/2882.

130 *Called Sputnik 1, the 22-inch-round aluminum ball weighed 183 pounds:* Roger D. Launius, "Sputnik and the Origins of the Space Age," https://history.nasa.gov/sputnik/sputorig.html#american.

131 *scientists all over the world, including the United States, had been working collaboratively toward the same goal:* Roger Launius, "Sputnik and the Origins of the Space Age."

132 *But in the meantime, another, unapproved satellite plan that had been developed by a team of US Army researchers for the IGY satellite effort resurfaced:* Roger Launius, "Sputnik and the Origins of the Space Age."

133 *He began working with congressional leaders to develop plans for a national agency dedicated to the exploration of space:* Roger Launius, "Sputnik and the Origins of the Space Age."

136 *That's when the president of Hampton Institute at the time, Alonzo G. Moron, stepped up and offered Mrs. Parks a job:* Dr. William Harvey, "Hampton University and Mrs. Rosa Parks: A Little Known History Fact," Hampton University, http://www.hamptonu.edu/news/hm/2013_02_rosa_parks.cfm.

136 *President Eisenhower signed into law the Civil Rights Act of 1957:* "Civil Rights Act of 1957, National Archives, Dwight D. Eisenhower Presidential Library, Museum & Boyhood Home, https://www.eisenhowerlibrary.gov/research/online-documents/civil-rights-act-1957.

chapter 8: LOVE WHAT YOU DO

144 *A poll by pioneer opinion researcher Claude E. Robinson, which was taken just after the Soviets sent both Sputniks into orbit:* National Archives, Dwight D. Eisenhower Presidential Library, https://www.eisenhowerlibrary.gov/sites/default/files/research/online-documents/sputnik/4-14-58.pdf.

144 *Just six days after the birth of NASA, a special panel, led by Langley assistant director Robert R. Gilruth:* Andrew Chaikin, "Bob Gilruth, the Quiet Force Behind Apollo," *Air & Space Magazine*, February 2016, https://www.airspacemag.com/history-of-flight/quiet-force-behind-apollo-180957788/.

146 *The Space Task Group decided to call the first space travelers "astronauts":* "Project Mercury Overview: Astronaut Selection," NASA, November 30, 2006, https://www.nasa.gov/mission_pages/mercury/missions/astronaut.html.

146 *NASA records show that 508 service records were examined, which narrowed the list to 110 men who met the criteria:* "Project Mercury Overview: Astronaut Selection," NASA.

149 *Virginia Governor Lindsay Almond shut down schools that were*

trying to comply with the US Supreme Court's order to desegregate: James H. Hershman Jr., "Massive Resistance," *Encyclopedia Virginia*, https://www.encyclopediavirginia.org/massive_resistance#start_entry.

152 *In 1962, students from North Carolina A&T and Bennett in Greensboro started a new round of protests, targeting about a dozen businesses that remained segregated:* Karen Hawkins and Cat McDowell, "Desegregation of Greensboro Businesses, 1962–1963," UNC Greensboro, http://libcdm1.uncg.edu/cdm/essay1963/collection/CivilRights.

153 *The president of Bennett, Dr. Willa Beatrice Player, was a fascinating woman who played a critical role during the protests:* "Profiles in Black History, *News & Record*, January 17, 2004, https://greensboro.com/life/profiles-in-black-history/article_fff9a7f3-8f8f-540b-8171-f9a216baa824.html.

154 *On May 17 and 18 that year, about seven hundred protesters, mostly students, were arrested:* "Mass Action in Greensboro (May–June)," African American Civil Rights Network, https://www.crmvet.org/tim/timhis63.htm.

155 *On January 31, 1961, our space team launched into space a chimpanzee:* "Mercury Redstone 2," *NASA space Science Data Coordinated Archive*, https://nssdc.gsfc.nasa.gov/nmc/spacecraft/display.action?id=MERCR2.

156 *Millions of television viewers tuned in:* Bob Granath, "Shepard's Mercury Flight Was First Step on the Long Journey to Mars," NASA's Kennedy Space Center, May 2, 2016, https://www.nasa.gov/feature/shepards-mercury-flight-was-first-step-on-the-long-journey-to-mars.

158 *The president asked Congress to approve a whopping seven billion to nine billion dollars:* "Space Program—President Kennedy's Challenge," John F. Kennedy Presidential Library and

Museum, https://www.jfklibrary.org/learn/about-jfk/jfk-in-history/space-program.

159 *A few years later, in the early 1950s, NACA purchased its first IBM computers, an IBM 604 Electronic Calculating Punch and the IBM 650:* Margot Shetterly, *Hidden Figures* (New York, HarperCollins Publishers, 2016), 138.

160 *Soviets sent yet another of its cosmonauts, Gherman Titov, into space on October 6, 1961:* Loyd S. Swenson Jr., James M. Grimwood, and Charles C. Alexander, *This New Ocean: A History of Project Mercury* (Washington, DC, NASA, Office of Technology Utilization, 1966), 378.

chapter 9: SHOOT FOR THE MOON

168 *The federal government had financed the program at colleges and flight schools throughout the country:* "Civilian Training Program," National Park Service, Tuskegee Airmen National Historic Site, https://www.nps.gov/tuai/civilian-pilot-training-program.htm.

169 *Tuskegee trained an estimated eleven hundred combat pilots:* "Training For War," National Park Service, American Visionaries exhibit, https://www.nps.gov/museum/exhibits/tuskegee/airwar.htm.

171 *The Birmingham Children's Crusade, as the demonstrations would come to be called, had been organized:* Kim Gilmore, "The Birmingham Children's Crusade of 1963," updated January 28, 2020, https://www.biography.com/news/black-history-birmingham-childrens-crusade-1963.

172 *Negro deejays helped spread the word:* "The Children's Crusade," African American Civil Rights Network, https://www.crmvet.org/tim/timhis63.htm#1963bham.

172 *The international embarrassment and public pressure pushed President John F. Kennedy and his brother Attorney General Robert Kennedy:* "Birmingham Campaign," The Martin Luther King, Jr. Research and Education Institute, Stanford University, https://kinginstitute.stanford.edu/encyclopedia/birmingham-campaign.

172 *By May 10, Dr. King and Rev. Shuttlesworth announced a major victory:* "Birmingham Campaign," The Martin Luther King, Jr. Research and Education Institute, Stanford University.

172 *a bomb ripped through the Gaston Motel:* "Birmingham Campaign," The Martin Luther King, Jr. Research and Education Institute, Stanford University.

173 *Governor George Wallace blocked the entrance to the University of Alabama's auditorium:* "Governor George C. Wallace's School House Door Speech," Alabama Department of Archives and History, updated December 12, 2012, https://archives.alabama.gov/govs_list/schooldoor.html.

173 *The Kennedy brothers again were forced to intervene, sending a hundred troops:* "June 11, 1963: From George Wallace to John Kennedy, a Momentous Day for Civil Rrights," Morgan Whitaker, MSNBC, June 11, 2013, https://www.msnbc.com/politicsnation/june-11-1963-george-wallace-john-ke-msna60336.

173 *The same evening in Jackson, Mississippi, Myrlie Evers and her children had watched President Kennedy's speech:* Jerry Mitchell, "'Memories' of home: What Does Myrlie Evers Have to Say about National Monument Designation," *Clarion Ledger*, March 13, 2019, https://www.clarionledger.com/story/news/local/2019/03/13/myrlie-evers-national-monument-home-medgar-evers-jackson-mississippi/3148021002/.

174 *It would become public in the years ahead that civil rights leader*

Anna Arnold Hedgeman, who was on the march's organizing committee: Arlisha Norwood, "March on Washington for Jobs and Freedom," National Women's History Museum, https://www.womenshistory.org/resources/general/march-washington-jobs-and-freedom.

176 *About 70 percent of Negroes had voted for President Kennedy:* "John F. Kennedy's Legacy Resides in African-American History," *Harvard Political Review*, December 1, 2013, https://harvardpolitics.com/john-f-kennedys-legacy-resides-african-american-history/.

177 *About seven hundred volunteers joined the Mississippi crusade:* "The Mississippi Summer Project," Public Broadcasting Service, https://www.pbs.org/wgbh/americanexperience/features/freedomsummer-project/.

178 *While doing so, the investigators found the bodies of eight other Negroes:* "Murder in Mississippi," Public Broadcasting Service, https://www.pbs.org/wgbh/americanexperience/features/freedomsummer-murder/.

181 *On May 16, 1963, astronaut Gordon Cooper flew on Project Mercury's final mission:* "Mercury Crewed Flights Summary," NASA, November 30, 2006, https://www.nasa.gov/mission_pages/mercury/missions/manned_flights.html.

182 *So the Mission Control Center operations moved from Cape Canaveral, Florida:* "1965: Mission Control Transfers to Houston," NASA, April 9, 2018, https://www.nasa.gov/mediacast/1965-mission-control-transfers-to-houston.

183 *Astronauts Virgil I. "Gus" Grissom (one of the original seven astronauts in Project Mercury), Edward H. White II, and Roger E. Chaffee climbed aboard the spacecraft . . . all three were dead:* Dr. David R. Williams, "The Apollo 1 Tragedy," NASA, https://nssdc.gsfc.nasa.gov/planetary/lunar/apollo1info.html.

183 *The main cause of the fire was determined to be electrical:* Dr. David R. Williams, "The Apollo 1 Tragedy."

chapter 10: FINISH STRONG

185 *Major Lawrence was among seventeen air force pilots chosen for a space venture, the Manned Orbiting Laboratory:* James Hill, "1st Black Astronaut Added to Space Hero Roster," *Chicago Tribune*, October 27, 1997, https://www.chicagotribune.com.

186 *The MOL program that had trained Major Lawrence for space was canceled in 1969:* John Uri"50 Years Ago: NASA Benefits from Manned Orbiting Laboratory Cancellation," NASA Johnson Space Center, June 10, 2019, https://www.nasa.gov/feature/50-years-ago-nasa-benefits-from-mol-cancellation.

186 *a brilliant young man, who not only had graduated from high school at age sixteen . . . classical pianist:* "Robert Lawrence: First African-American Astronaut," NASA, updated February 21, 2018, https://www.nasa.gov/feature/robert-lawrence-first-african-american-astronaut.

187 *Dwight was a promising U.S. Air Force pilot with an aeronautical engineering degree from Arizona State University:* Charles L. Sanders, "The Troubles of 'Astronaut' Edward Dwight," *Ebony*, June 1965, https://books.google.com/books?id=Nd4DAAAAMBAJ&pg=PA29#v=onepage&q&f=falseEric.

192 *The campaign's goal was to draw attention to the poverty and suffering in America's cities:* Eric Niiler, "Why Civil Rights Activists Protested the Moon Landing," updated July 18, 2019, https://www.history.com/news/apollo-11-moon-landing-launch-protests.

193 *the huge Saturn V rocket—as tall as a 36-story building:* "Saturn V," NASA, updated September 16, 2011, https://www.nasa.gov/centers/johnson/rocketpark/saturn_v.html.

197 *They also took photographs:* "Apollo 11 Mission Overview," updated May 15, 2019, https://www.nasa.gov/mission_pages/apollo/missions/apollo11.html.

chapter 11: LAND ON A FIRM FOUNDATION

203 *But our excitement quickly turned to horror when just seventy-three seconds into the launch . . .:* Howard Berkes, "Challenger: Reporting a Disaster's Cold, Hard Facts," National Public Radio, January 28, 2006, https://www.npr.org/2006/01/28/5175151/challenger-reporting-a-disasters-cold-hard-facts.

204 *Astronaut Group 8, which included thirty-five space travelers:* "1978 Astronaut Class," NASA, updated August 6, 2017, https://www.nasa.gov/image-feature/1978-astronaut-class.